Doug Arent's story is indeed remarkable: He takes us beyond the despair and gloom of climate change to a world of abundance and possibility. This isn't a fictional story but a way of thinking that combines technology, economics, policy and new ways of creating knowledge itself.

Richard Kauffman
Chair, Generate Capital

* * *

Doug Arent has actively sought to understand and help to shape the nation's clean energy transition and has been both an on-field player and occupant of a front-row seat for the transformation of technologies from dreams to reality. *Our Renewable Energy Future* brings a unique vantage point to explain key elements of that transformation in ways accessible to the interested and reasonably informed layperson, with analytic rigor and expert insights based on decades of work. This book is a must-read for anyone wanting to know where we've come and where we're going in our renewable energy transition.

Sue Tierney
Senior Advisor, Analysis Group
Former Assistant Secretary for Policy, U.S. Department of Energy
Chair of Board Resources for the Future and Vice Chair of the Board
World Resources Institute

* * *

Doug Arent offers us a remarkable perspective on the (r)evolution of our energy system, now in full swing. A truly inspiring read, leaving us with renewed optimism and excitement that a sustainable and renewable future is now within reach.

Vincent Petit
Senior Vice President, Energy Transition
Sustainability Research Institute, Schneider Electric

* * *

A captivating and well-written history and future of renewable energy. A must-read book for all interested in a visionary view of our common energy future, what would need to be done and how this would be possible.

Prof. Dr. Dr.h.c. Nebojsa Nakicenovic
Vice Chair, Group of Chief Scientific Advisors, European Commission
Distinguished Emeritus Scholar and Former Deputy Director General, IIASA
Former Professor of Energy Economics, Vienna University of Technology

* * *

Doug Arent is the perfect person to have written such an exceptional book on our renewable energy future. His leadership at the National Renewable Energy Laboratory and his prominent role in prestigious think tanks dealing with this subject have enabled him to provide insight well beyond technological developments into policy, economic considerations and even geopolitics. It is a spectacular read!

Clint Vince, Esq.
Chair, Energy, Dentons US LLP

* * *

This is the right book at the right time for our collective energy and climate challenges, and Doug is the right person to have pulled it together. It is simultaneously optimistic and realistic and required reading for those of us who want to make the energy system better.

Michael Webber
Josey Centennial Professor in Energy Resources
University of Texas at Austin

* * *

We are witnessing an historic transition from fossil-fuel-based energy to a new world of renewables, driven by a surge in technology, the demand for energy security and environmental pressure. The shift, while inexorable, can come none too soon: without it, we will be subject to the increasingly dire consequences of global warming, which will only be forestalled with a quick transition. In this comprehensive and masterfully written book, replete with details and examples, Doug Arent presents the history of his field from the perspective of someone who has not only lived through it but helped drive it. It is a valuable read for both the informed policymaker who must understand the history of the sector to effectively shape its future, as well as the lay reader who seeks to be informed about this most fundamental driver of our world: the energy system and its renewable transition.

Jonathan Pershing
Former U.S. Special Envoy for Climate Change

* * *

With renewable energy capacity globally set to continue its near-exponential growth, there is need for a more holistic understanding of the evolution of clean energy technology and our future energy system. In his new book *Our Renewable Energy Future: The Remarkable Story of How Renewable Energy Will Become the Basis for Our Lives*, Arent demonstrates his flair for systems thinking by explaining, easily and accessibly, the complex interaction of technology, economics, social dynamics, policy and geopolitics shaping the transition to a clean energy future. A timely book for anyone interested in a world powered by more renewable energy.

Lawrence Jones
Senior Vice President, EEI
Editor, Renewable Energy Integration *(1st and 2nd editions)*

* * *

Our Renewable Energy Future is an exceptional book that encapsulates the collaborative evolution of this sector, from Dr. Elliot Berman's Exxon-supported solar cell innovation to today's global industry, including the Internet of Things and Blockchain. It highlights the crucial roles of academia, entrepreneurs, financiers, regulators and policy makers in this organic growth. Balancing intricate technical discussions with real-world experiences, Doug Arent's book provides a compelling analytical framework and rich storytelling. Its insightful policy recommendations make it a must-read guide for policy makers to understand and contribute to a net-zero world. This book is a testament to the power of collective endeavor in realizing the potential of renewable energy, serving as an inspirational blueprint for future advancements.

Leonardo Beltran
Former Deputy Secretary of Planning and Energy Transition
Government of Mexico
Distinguished Visiting Fellow
Center on Global Energy Policy, Columbia University

* * *

Doug Arent is a global leader advancing gigaton solutions to clean energy deployment. He is the consummate collaborator, identifying technologies when they show promise and joining them with market solutions for deployment. His abilities to create action among key players, institutions and new actors is legend. Doug foments creativity and responsiveness. His opinions influence capital, corporations and governments. His book capturing insights and strategies from decades at the center of clean energy and development is a must-read for those who believe that speed and scale matter, and that our planet and people will all be better as a result of our renewable energy future.

Jeff Weiss
Executive Chairman, Distributed Sun and truCurrent

* * *

A wealth of thoughtful perspectives and key information on the birth and growth of modern renewable energy from a top player who has seen it all. *Our Renewable Energy Future* provides critical insights into the coming decades and how renewable energy might finally realize its massive potential to address the climate crisis and advance the economy for people and the planet.

Dan Reicher

Former U.S. Assistant Secretary of Energy
for Energy Efficiency and Renewable Energy
Former Director of Climate and Energy Initiatives, Google
Senior Scholar, Stanford Doerr School of Sustainability

Our Renewable Energy Future

The Remarkable Story of How
Renewable Energy Will Become
the Basis for Our Lives

Our Renewable Energy Future

*The Remarkable Story of How
Renewable Energy Will Become
the Basis for Our Lives*

DOUGLAS ARENT *Ph.D.*

National Renewable Energy Laboratory, USA

World Scientific

NEW JERSEY · LONDON · SINGAPORE · BEIJING · SHANGHAI · HONG KONG · TAIPEI · CHENNAI · TOKYO

Published by

World Scientific Publishing Europe Ltd.

57 Shelton Street, Covent Garden, London WC2H 9HE

Head office: 5 Toh Tuck Link, Singapore 596224

USA office: 27 Warren Street, Suite 401-402, Hackensack, NJ 07601

Library of Congress Control Number: 2023057769

British Library Cataloguing-in-Publication Data
A catalogue record for this book is available from the British Library.

Cover artwork by Mekenna Arent

OUR RENEWABLE ENERGY FUTURE:
THE REMARKABLE STORY OF HOW RENEWABLE ENERGY WILL BECOME THE
BASIS FOR OUR LIVES

ISBN 978-1-80061-493-2 (hardcover)
ISBN 978-1-80061-606-6 (paperback)
ISBN 978-1-80061-494-9 (ebook for institutions)
ISBN 978-1-80061-495-6 (ebook for individuals)

For any available supplementary material, please visit
https://www.worldscientific.com/worldscibooks/10.1142/Q0441#t=suppl

Desk Editor: Nimal Koliyat/Rosie Williamson/Shi Ying Koe

Typeset by Stallion Press
Email: enquiries@stallionpress.com

For Etienne and Mekenna,
And their children,
And all our children.

This book and the journey behind it would not be possible without
the love and support of Joan Whalen Arent.
My love and many thanks.

Preamble and Thesis

This book focuses on clean energy, technology evolution, and the direction in which our energy system is headed. It builds from a life of passion focused on energy, sustainable development, and in particular renewable energy technologies. The journey started with a passion for math and science. In the early 1970s, the U.S. and the world were facing an oil crisis. The U.S. responded with a series of actions, including the creation of what was then called the Solar Energy Research Institute. At that time, I was merely in high school. But, the idea of a renewable energy future struck me. I wrote a paper on the potential of solar energy and still have a copy of it, which you can see in the appendices. What follows in the pages of this book is the evolutionary journey of technological, economic, policy, knowledge and social and institutional factors over the last 40 years. We will trace the arc of recent technological innovations and synthesize innovations across multiple interacting industries into a description of Our Renewable Energy Future.

From a personal perspective, I have been honored to be a citizen of the world and have traveled widely, having focused on energy for development and clean-energy transitions. With deep gratitude, I have interacted with many leading scientists, policymakers, and thought leaders over the past decades. This includes ongoing efforts, conversations, and debates with Amory Lovins, the founder of the Rocky Mountain Institute (now RMI), and Robert Fri, who was the deputy administrator and then director of the Energy Resources and Development Agency (ERDA), in the mid-1970s, with whom I have had the honor of serving on a National Academies Panel. ERDA was the predecessor of what is now the U.S. Department of Energy.[1] I have also been honored to serve with leading scientists at the IPCC, including my good friend and colleague Nebojsa Nakicenovic (Naki), who worked for years at the International Institute for Applied Systems Analysis (IIASA), in Austria. IIASA was established after World War II as a collaborative scientific leading institution advising governments and the United Nations. I am also indebted to a small cohort of renewable energy thought leaders and analysts, including Paolo Frankl, head of the Renewable Energy Division at the International Energy Agency, who have formed a collaborative community to share knowledge, advance understanding, and advise governments and corporate leaders over the decades. Many other have influenced my understanding and career, including hundreds of scientists and scholars at National Renewable Energy Laboratory (NREL). I am deeply indebted to them for pushing the boundaries of technology, and understanding energy modeling, policy, and economics.

It is not only worth noting but also disappointing to recognize that, over the past 40 years, we have made measurable progress for those without access to modern energy (nearly a billion people in the last few decades) and all that it affords in terms of education, clean water, food, and economic prosperity, but still nearly 1 billion people remain without access to modern energy. This challenge offers reflection on how much opportunity and obligation we have yet to fulfill.

History offers valuable insights. In 2022, with the Russian invasion of Ukraine and the use of energy as a strategic, economic, and political weapon, reflections from the oil embargo of the 1970s and prior "energy crises" (which were in fact the result of political actions, not a lack of molecules) offer a stark reminder of the potential role of energy as a weapon and the enhanced recognition of its critical role in life. Further, there is now a fresh recognition of supply chain concentrations and potential geopolitical and economic risks. This reset of geopolitical perspectives is a fundamental reset of attitudes toward globalization, nimble on-demand supply chains, geopolitical security, and national interests. The prior few decades were dominated by a set of assumptions that globalization was a positive social good and that all nations would prosper if there were deeply intertwined supply chains that would support global economic prosperity. Today, it is clear that globalization is being rethought both from a geopolitical standpoint and also from an economic standpoint, by both policymakers and corporate leaders. There is considerable attention being paid and momentum being built to reset the global supply chain dynamics. While it will take decades to find a new, metastable equilibrium, it is fairly clear that the world will develop a more polycentric approach rather than a relatively uniform approach of least cost economics without deeper risk-adjusted geopolitical and diversity dynamics being factored in. As one CEO recently said, if we had managed our supply chain so naïvely with so much risk concentration, we would have been fired. Anticipation of quick answers and a reset of very large global supply chains will need to be muted given time-constants for development, infrastructure investment, labor restructuring, etc.

From a climate mitigation perspective, what we are facing, and what we must realize, is a fundamental transformation of our energy system toward one that does not emit greenhouse gases or have detrimental environmental impacts and assures a just and fair energy transition worldwide. Energy, whilst a connective currency across all of our ecosystems and critical for the quality of life that so many of us have come to enjoy and take for granted, is only one of our critical infrastructures. We must think about water, land use, the oceans, and, of course, people.

I am reminded of early work that was done on Small Solutions or Limits to Growth (which had its 40th anniversary in 2022). These were fundamental concepts by early thinkers who recognized that an engineered resource-intensive, highly consumptive, and throwaway society would have detrimental impacts on our earth and our lives. It is only now that the bulk of the world is recognizing this and moving toward hopefully a much more sustainable future. A few courageous leaders have embraced new approaches to advancing sustainable, scalable solutions, such as Paul Polman, former CEO of Unilever, who for years worked with Jeff Seabright, the chief sustainability officer (CSO), to create a paradigm shift for net positive companies – those that give more than they take to employees, communities, the planet, and partners. I sincerely hope all leaders will embrace this paradigm expeditiously.

Prologue

It is 2050. The world is free of energy poverty, and there are sufficient energy services for a decent life for everyone. There is clean water, electricity, fast digital communications, well-insulated and comfortable housing, piped water and sanitation, and abundantly productive agriculture which provides enough food to feed everyone a comfortable and healthy diet. And, we are doing it all, everywhere, without polluting the earth, the oceans, or the atmosphere. How has this come to be? We have learned to harness renewable energy sources around the world, combined with other low-carbon solutions, enabled by a complex mix of digital controls, distributed energy systems, integrated energy systems, and interconnected power grids all transmitting clean electrons – the backbone of our clean-energy economy. Those clean electrons are used to create and support clean chemicals and fuels, which in turn go into nearly every product that we so enjoy. Steel, aluminum, and concrete are all made via processes that leverage this renewable electricity-dominated future. Plastics – and nearly all consumable materials – come from a full reuse and recycle industry. There is no more plastic waste. Energy is not a limitation, but an enabler of economic advancement, education, health…

Utopia? Perhaps, or perhaps not.

About the Author

Douglas Arent has conducted research on energy and sustainability for more than 40 years, publishing extensively on topics such as clean energy, renewable energy, power systems, natural gas, and the intersection of science and public policy. Since 2020, Dr. Arent has been the Executive Director of Strategic Public–Private Partnerships at the National Renewable Energy Laboratory in Golden, Colorado, U.S. At NREL, he spent more than 20 years building and leading the Strategic Energy Analysis team. He previously served as the Deputy Associate Laboratory Director of the Scientific Computing and Energy Analysis Directorate, the Founding Executive Director of the Joint Institute for Strategic Energy Analysis, and the Director of the Strategic Energy Analysis Center and the Integrated Applications Center. Additionally, Dr. Arent is a Distinguished Fellow of the World Economic Forum and a Senior Visiting Fellow at the Center for Strategic and International Studies. He is a member of the Keystone Energy Board and the Global Advisory Board for the Yale Emerging Climate Leaders Fellowship. Dr. Arent is also an External Fellow at the Center on Global Energy Policy at Columbia University School of International and Public Affairs and the Andlinger Center for Energy and the Environment at Princeton University. He is the Editor-in-Chief of *Renewable Energy Focus* and is on the editorial board of the journals *Energy Policy* and *Renewable and Sustainable Energy Reviews*. Previously, Dr. Arent served on the Steering Committee of the American Academy of Arts and Sciences on Social Science and the Alternative Energy Future and on the National Research Council Committee to Advise the U.S. Global Change Research Program. He also served on advisory boards for the Post-Carbon Transition Project at the Institute for New Economic Thinking at Oxford University, U.K., and the Energy Academy of

Europe, the Netherlands. Dr. Arent was the coordinating lead author for the Fifth Assessment Report of the Intergovernmental Panel on Climate Change. He has been a member of the Policy Subcommittee of the National Petroleum Council Study on Prudent Development of North America Natural Gas and Oil Resources, the National Academy of Sciences Panel on Limiting the Magnitude of Future Climate Change (2008–2010), and the Executive Council of the U.S. Association of Energy Economists. Dr. Arent has a Doctorate in Physical Analytical Chemistry from Princeton University, a Master of Business Administration degree from Regis University, and a Bachelor of Science degree from Harvey Mudd College.

Acknowledgments

This book would not have been possible without the innovative creativity of hundreds of colleagues at the National Renewable Energy Laboratory and around the world. I am honored to be a part of that creative community and the broader community which has been addressing climate change, clean energy, and environmental sustainability for the past several decades.

With many thanks to the Rockefeller Foundation for providing a Residency Scholarship during which substantial progress was made on this book.

Contents

List of Figures

❧

Chapter 4

Chapter 5

Abbreviations

°C	Degrees celsius
5G	5th generation (cellular technology)
AC	Alternating current
AGC	Auto generation control
AI	Artificial intelligence
AR3	Assessment report 3
AR5	Assessment report 5
AR6	Assessment report 6
ASTM	American Society for Testing and Materials
BTU	British thermal unit
CAES	Compressed air energy storage
CCS	Carbon capture and storage
CCUS	Carbon capture, utilization, and storage
CEAP	Circular economy action plan
CF	Capacity factor
CGE	Computational general equilibrium
CO_2	Carbon dioxide
COP	Conference of parties
CSP	Concentrating solar power
DERs	Distributed energy resources
EIA	U.S. Energy Information Administration
EPRI	Electric Power Research Institute
ERDA	Energy Research and Deployment Agency
ESG	Environmental, social, and governance
ET	Energy technology
ETS	Emissions trading system
EV	Electric vehicle

EVTOL	Electric vertical take-off and landing vehicle
FIT	Feed-in tariff
GDP	Gross domestic product
GHG	Greenhouse gas
GNP	Gross national product
GPS	Global positioning system
GW	Gigawatt
GWEC	Global Wind Energy Council
H_2	Hydrogen
H_2O	Water
HEMS	Home energy management system
HOMER	Hybrid Optimization of Multiple Energy Resources
IEA	International Energy Agency
IEEE	Institute of Electrical and Electronics Engineers
IIASA	International Institute of Applied Systems Analysis
IIJA	Infrastructure, Investment, and Jobs Act
IoET	Internet of Energy Things
IPCC	Intergovernmental Panel on Climate Change
IRA	Inflation Reduction Act
IRENA	International Renewable Energy Agency
IRP	Integrated resource plan
IT	Information technology
ITC	Investment tax credit
kg	Kilogram
km	Kilometer
kw	Kilowatt
kWhr	Kilowatt hour
LADWP	Los Angeles Department of Water and Power
LCOE	Levelized cost of energy
LED	Light-emitting diode
Li	Lithium
LIB	Lithium-ion battery
LNG	Liquified natural gas

MBD	Million barrels per day
ML	Machine learning
MT	Million tons
MtCO$_2$ eq	Million metric tons of CO$_2$ equivalent
MW	Megawatt
MWac	Megawatt, alternating current
MWhr	Megawatt hour
N$_2$	Nitrogen
NDC	Nationally determined contribution
NEMS	National energy modeling system
NO$_x$	Nitrous oxides
NREL	National Renewable Energy Laboratory
O$_2$	Oxygen
PACE	Platform for a Circular Economy
PES	Power and Engineering Society
PPA	Power purchase agreement
PTC	Production tax credit
PURPA	Public Utility Regulatory Policies Act
PV	Photovoltaic
R&D	Research and development
RDD&D	Research, development, demonstration, and deployment
RE	Renewable energy (energies)
RECs	Renewable energy certificates
ReEDS	Regional energy deployment system model
REF	Renewable electricity futures
REIPP	Renewable energy-independent power production
ReOpt	Renewable energy optimization
RMI	Rocky Mountain Institute
RPS	Renewable portfolio standard
SAFs	Sustainable aviation fuels
SDGs	Sustainable development goals
SERI	Solar Energy Research Institute
SO$_x$	Sulfur oxides

TB	Terabyte, or 10^{15} bytes
TCF	Trillion cubic feet
TW	Terawatt
TWhr	Terawatt hour
UNFCCC	United Nations Framework Convention on Climate Change
W	Watt
Wdc	Watts in direct current
WinDS	Wind deployment system model

Introduction

This book focuses on the history of clean energy, its future direction, and our transition to sustainable energy economies. It is written for an audience that understands technology, but more importantly seeks to understand the interplay between technology and society. It is not written as a technical textbook, nor does it cover all energy – it focuses on renewable electric technologies, their advances, and the technical, social, institutional, policy and financial innovations enabling them to dominate our energy future. It will touch on the role of nuclear energy; the ongoing use of fossil fuels and the technologies to decarbonize those energy sources; alternative technologies for carbon dioxide (CO_2) capture, utilization, and storage; and the role of hydrogen, clean fuels, and chemicals. It is written as a book that synthesizes the factors that impact a technology's ability to address individual, institutional, and social needs. The book is broken down into four primary phases of the Renewable Energy Revolution (some may say evolution, as it has been occurring over decades, but energy infrastructure is both capital intensive and long lived – thus, a few decades can be considered a "revolution" as such). The Phases are as follows:

Phase I: *Promises* – The Technologies of Tomorrow (1950s–1980s)

Phase II: *Inching Along* – Early Commercialization (1990s–2000)

Phase III: *From Problem Child to Problem Solver* – The Era of Exponential Growth (2000–2020)

Phase IV: **Dominance** – From Petro to Electro (2020+)

and

Phase V: *Beyond 2050* – A Global Net-Zero Energy Economy (2050+)

Further, the book will use a framing that evaluates six key factors throughout the four primary phases. The six factors are as follows:

1. **Sustainable Technology Capabilities:** This refers to the engineering and technical capabilities of a given technology. Technology here is actually meant to be quite inclusive, covering, for example, not only the fundamental provision of creating power but also the ability to integrate that power into a complex system. Integration includes not only simply provision but also flexibility, dynamic management, and provision of multiple different capabilities.

2. **Economics/Finance:** Here, the fundamental cost structures of these technologies are incorporated, but also included are considerations of how business models change their affordability, moving from a capital purchase, for example, to leasing or just an operating payment. The finance side includes the full perspective of the finance industry, stemming from government enabling conditions through to public and private capital markets, project financing, banking, digital financing, etc.

3. **Policy/Regulatory (Local, National, and Global):** This perspective includes the complex combination of carrots and sticks from a policy perspective, also covering multiple jurisdictions which may impact a given suite of technologies in a given location.

4. **Knowledge (Models, Tools & Data for Planning & Operations and Business Models):** Here, the collective set of intelligence that is gained within the academic, scientific, engineering, and corporate communities is reflected upon. It also extends beyond that, including knowledge gained around social inclusion, processes, business models, communications, and other knowledge critical for, or inhibiting, a given technical solution to fit into and prosper in a given societal context.

5. **Social Willingness:** This factor incorporates the collective use of the bulk of society. This section will not be diving into survey information, generational gaps, or extreme positions, but representing what I believe is a majority view of clean energy and environmental sustainability.

6. **Institutional Willingness/Political Economy Dynamics:** This factor differs from the previous five as it reflects the perspective among institutions

toward either supporting or inhibiting clean-energy transition and advancing environmental stewardship.

Why are those six called the framing factors? Based on nearly 40 years of experience working both as a scientist and a technologist in economics and policy spheres around the world, it is clear the technology alone is insufficient to address the needs of the individual or society. Technology must meet or exceed a service expectation, at a price, and at a service level that each of us cares about and is willing to pay for. Additionally, technology must align with the social, political, and economic current conditions and contexts. The solutions must scale appropriately and of course be profitable to the entrepreneurs and companies who have invested their time and financial resources. They must be attractive to investors, and they must resonate with social willingness and the license to operate.

Over the past many decades, while it has been known and thought about by a few energy experts that the world has limits to extractive industries (not a lack of molecules, but above-ground issues) and polluting processes, leaders and millions of others have only recently woken up to the fact that the world's climate is indeed in a crisis. While I have been focused on climate mitigation and the potential of clean-energy technologies for over four decades, this recent awakening by the world's leaders (government, non-government, and corporate) and our youth who are demanding change has finally built sufficient momentum, in combination with technological progress, allowing us to foresee a pathway to a sustainable energy future. That sustainable energy future is a fundamental requirement for economic prosperity and true sustainability as laid out by the Bruntland Commission in 1987:

Sustainable solutions meet the needs of the present without compromising the ability of future generations to meet their own needs.

Thus, the book will reflect on history as it gives us perspective, but more importantly it will look forward and map out what our sustainable energy future looks like, in particular Phase IV, where renewable energy moves from the mainstream to domination of our energy economies and the foundation of our broader economic prosperity.

Chapter 1
Phase I: *Promises* – The Technologies of Tomorrow (1950s–1980s)

Key Themes

- Pioneers

- Inventors

- Visionaries

- Promise of the future

Technology "Discovery" and *Very* Early Niche Markets

- High-value markets

- Expensive and uncompetitive with other energy options

- Low performance

Factors

A. Sustainable technology capabilities
B. Economics/Finance
C. Policy/Regulatory (Local, national, and global)
D. Knowledge (Models, tools & data for planning & operations and business models)
E. Social willingness
F. Institutional willingness/Political economy dynamics

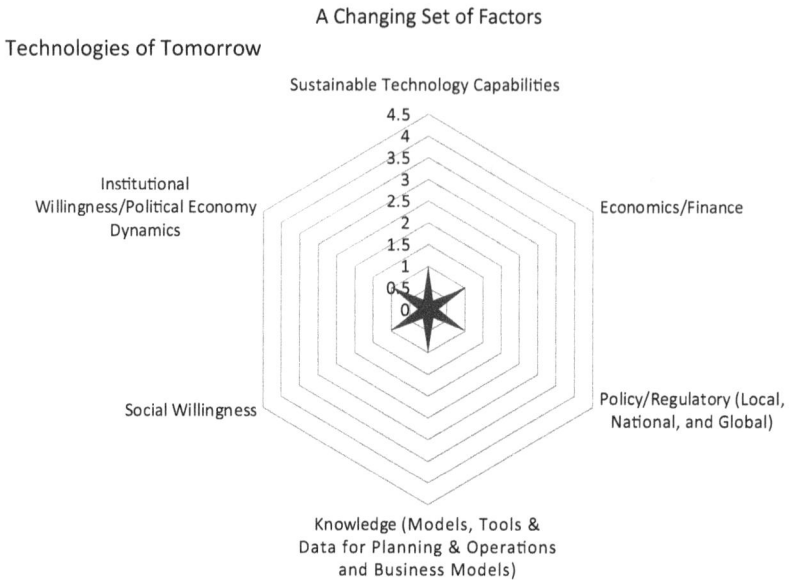

A Changing Set of Factors

Technologies of Tomorrow

Sustainable Technology Capabilities

Institutional
Willingness/Political Economy
Dynamics

Economics/Finance

Social Willingness

Policy/Regulatory (Local,
National, and Global)

Knowledge (Models, Tools &
Data for Planning & Operations
and Business Models)

Figure 1. A changing set of factors for Phase I.

As a short-hand visual, Figure 1 describes the relative breadth and intensity of the various factors with higher numbers being more conducive to the expansion and deployment of renewable energy technologies. The breadth is indicative of the range of countries which are engaged and actively supporting clean-energy technologies and in particular renewable energies. This is meant to be a notional guide to give a sense of the state of the different metrics in the Phases.

Phase I was predominantly focused on technology development through the mid-1970s. This was the time of the invention of the first practical solar cell, based on the photoelectric effect that has been known for over 100 years.[2] It was in the Seventies, as a reaction to the global oil crisis, that deployment policies began to be developed. It was also the time of early expansion of wind power for electricity generation. This was epitomized by the growth of wind farms in Southern California with turbines of, in today's terms, a relatively small size (e.g., 100s of kW, vs the MW-scale turbines of today). Of course, wind power itself has been harnessed for many thousands of years to grind grain and even produce or move water. Water power, of course, has also been around for centuries.

More significantly perhaps, during this time, our awareness of technological progress was spurred by a combination of rapid media attention, e.g., television and radio, and the technology revolution inspired by the space program, sending a man to the moon, the expansion of air travel, and the modernization of our lives with more electronics.

For many, this was a time of inspiration, for example, early calculations and understanding of the enormity of solar energy reaching the earth. More energy reaches the earth via solar energy in a minute than the human and industrial systems consume in a year. For a sense of scale, 173,000 TW of energy reaches the earth continuously, greater than our annual primary energy consumption of approximately 167,000 TW hr per year. This fact inspired early solar scientists to strive for mechanisms to effectively harness this energy. The calculation, however, only looked at industrially harnessed energy. There is broader ecosystem science that of course recognizes the value and contributions of solar energy to the world's ecosystems. This includes temperature differentials that generate the winds, photosynthesis, and biological activities. However, within the pure industrial energy ecosystem complex, this recognition of the enormity of solar potential led to the creation of a vision for a solar future. That future, however, came from solar scientists. They were reflecting on the ideal. They knew little of how power systems worked and the challenges of introducing inverter-based renewable energy resources.[a] Much of this scientific knowledge and innovation would come in the following decades.

Reflecting on the factors that will be discussed throughout this book, one can see that nearly all of them are nascent at the very beginning, during Phase I. That is, when technology was just being envisioned or invented, the economics were uncompetitive without explicit policy support. For example, power from solar cells in 1970 cost approximately \$110/W (or \$640/MWhr (\$0.64/kWhr) vs average electricity prices of \$0.03/kWhr (nominal) in the U.S.). Wind power cost approximately \$250/MWhr in 1980 (2010 dollars).

Initially, there was little if any policy and regulatory support in any country. The oil crisis of the mid-1970s, however, did bring early policy interest to supporting

---

a Inverters are devices which convert direct current (DC) power that is produced by solar cells to alternating current (AC) power.

alternative energy technologies. These were early experiments for policy approaches, including state (predominantly in California) and federal tax credits in the U.S. for solar hot water, and early approaches to supporting distributed energy technologies, for example, in California where policymakers introduced the first feed-in tariff for solar energy systems. Building on the 1978 Public Utility Regulatory Policies Act (PURPA), California continued to innovate with policy support, including power purchase agreements, which allowed contracts between independent power producers and utilities, and tax credits. The combination of federal and state policies led to early expansion of solar and wind markets in California. This included the very successful early concentrating solar plants (CSP) in Southern California. Those plants utilized Fresnel lens concentrators with hot oil, included storage tanks and a steam turbine. This combination of solar plus storage to deliver power when needed from the grid became a much more important enabling feature in the future decades. These early pioneers laid a solid foundation for renewable technologies that could be viably integrated into the grid. What they faced was a very complex competitive pricing environment. For those in the wind industry, the technology was not as hardened as they had hoped. Many of the early installations in Southern California experienced technical failures. This gave wind a relatively poor reputation as a viable source of electric power generation for many years.

While there were commercial technologies and policy incentives to attract investment and entrepreneurs, the institutional processes and procedures proved immature in terms of providing durable solutions. Beyond wind turbines that failed, hot water systems had few if any quality standards. There were no installation standards or testing and verification standards. Many if not all of these became relics of the past. Costs for renewables, even with the fiscal policy incentives, were challenging given comparative costs of fossil fuel power generation at that time. Nuclear energy was just becoming commercial and was "too cheap to meter." Interestingly, at this phase in the U.S. in particular, there was a belief that natural gas was a limited resource and had to be reserved for uses other than power generation. This was codified into the Natural Gas Policy Act of 1978 which prevented natural gas from being used in power generation.

Knowledge, of course, was relatively restricted to the science community, and the public understood very little about it. Through the 1980s, solar scientists were creating solar cells, and at best they were able to put together small power systems with batteries that went onto space flights. The early space program also drove early innovations in hydrogen production, storage, and use, but not with the intent to spur competitive marketplace adoption. In 1963, Japan installed a 242-watt photovoltaic array on a lighthouse, the world's largest array at that time. In the 1970s, Dr. Elliot Berman, with help from Exxon Corporation, found a way to use less pure silicon to create a less costly solar cell, opening up opportunities for terrestrial applications. Solar cells began to power navigation warning lights and horns on many offshore gas and oil rigs, lighthouses, and railroad crossings, and domestic solar applications began to be viewed as sensible applications in remote locations where grid-connected utilities could not be extended affordably. Other off-grid applications focused on vaccine refrigeration, room lighting, medical clinic lighting, telecommunications, water pumping, grain milling, and classroom television.

Technology was, in today's view, relatively crude. It was only in 1977 when total photovoltaic manufacturing production reached 500 kW, but it took only another 5 years to reach 9.3 MW.

In the early part of Phase I, throughout the 1950s–1970s, social willingness was nonexistent as there was almost no social consciousness of clean-energy technologies and only a little comprehension or awareness of climate change. A few pioneers were publishing papers on our energy future and the need and opportunity for a different path. My long-time colleague Amory Lovins, founder of the Rocky Mountain Institute (now RMI), published his early work "Energy Strategy: The Road Not Taken?" in 1976 in which he articulated arguments for following "soft" energy solutions based on scalable renewable energy technologies, energy efficiency, and efficient mass transit.[3] Interestingly, he predicted that water would be one cause of the world's next resource crisis – which is now the case in many countries given water table extraction, pollution, and climate change.

At the broader institutional, political, and economic levels, there was little attention being paid to renewable technologies until the 1970s. In the U.S., the government initiated the creation of a national laboratory system through the Manhattan Project in the 1940s. This was brought on by World War II and the invention and creation of nuclear weapons systems. It was not until 1977 when the U.S. government created the U.S. Department of Energy. The department itself had, and still maintains, a broad mission, including national security, nuclear weapons and nuclear fuel management, energy-related basic sciences, comprising building the national science capability for accelerators, high-performance computing, and eventually programs. It was not until 1976, during President Carter's era and in the midst of the oil crisis, that Congress approved the creation of the Solar Energy Research Institute (SERI). Carter explained when he visited SERI in 1978, "We must begin the long, slow job of winning back our economic independence. Nobody can embargo sunlight. No cartel controls the sun. Its energy will not run out."

SERI is now the National Renewable Energy Laboratory (NREL), elevated in 1991 to the status of a national laboratory.

By the end of the 1980s, renewables had begun to blossom, albeit at a very small scale, e.g., 100s of MW, not GWs. Technologies were advancing at a moderately slow pace. Cost economics were uncompetitive but yet sufficient with fiscal incentives to attract investment. A few iconic installations became poster children for success, such as those in Barstow and Daggett, California. The technologies held the promise of the future, but were deployed in such small quantities that they had very little impact on the overall energy system and had not built sufficient momentum to continue to grow without substantial policy support.

A Synopsis of Early Policy and Regulatory Approaches and Milestones for Clean-Energy Technologies

1976: California established 10% investment tax credit for installation of renewable technologies, mainly solar energy.

1977: SERI was created (later became NREL).

1978: California's investment credit was increased to 55% and extended to wind energy until 1986 and to other alternative energy sources in the 1990s.

(Continued)

1978: California set a goal of the installation of 500 MW of wind capacity by the mid-1980s with the Wind Energy Program.

1978: The National Energy Act, including PURPA, was passed. This sought to improve and develop alternative energy sources. It required utility companies to buy electricity from qualified facilities that provided alternative methods of producing electric power. The purchasing cost was left to state regulation. California used anticipated high prices for natural gas and oil in their avoided cost calculations, enabling an attractive investment environment for renewable power.

1978: The Natural Gas Policy Act was enacted, which prohibited natural gas from being used in power generation due to the perception of it being a limited resource. This led to coal being promoted along with the advancement of nuclear generation.

1980s: California started several demonstration projects for wind energy.

1982: California's purchase cost for wind power dropped significantly after returning to the actual avoided cost. California's Public Utilities Commission (CPUC) created 10-year contracts that agreed on a power purchase rate of 6.9 cents/kWhr for both utility companies and qualifying alternative energy facilities.

1982: The worldwide production of photovoltaic cells exceeded 9.3 MW.

1983: The worldwide production of photovoltaic cells exceeded 21.3 MW, with sales of more than $250 million.

1983: Iowa adopted the world's first renewable portfolio standard (RPS), with an alternative energy production target of 105 MW from renewable

(Continued)

(Continued)

electricity sources in the state by the end of 1990 (more than 12,000 MW has been installed as of today).

1985: California installed 1,000 MW of wind capacity.

1986: The world's largest solar thermal facility, located in Kramer Junction, California, was commissioned. The solar field contained rows of mirrors that concentrated the sun's energy onto a system of pipes circulating a heat transfer fluid. The heat transfer fluid was used to produce steam.

1986: The Tax Reform Act was enacted, which extended tax credits for businesses for solar, ocean thermal, geothermal, and biomass energy generation.

1987: Fuel Use Act was repealed allowing natural gas to be used for power generation.

1990: California installed 1,799 MW of wind capacity.

1990: Amendments were made to the Clean Air Act (CAA), which led to the introduction of a new emission reduction program.

Chapter 2
Phase II: *Inching Along* – Early Commercialization (1990s–2000)

❧

Key Themes

- Niche markets to early adopters
- Expansion of approaches for the development of energy
- Further policy experiments
    - Feed-in Tariffs (FITs): Expansion in Europe
- Early integration of renewables: myths, experiments, and lessons

Factors

A. Sustainable technology capabilities
B. Economics/Finance
C. Policy/Regulatory (Local, national, and global)
D. Knowledge (Models, tools & data for planning & operations and business models)
E. Social willingness
F. Institutional willingness/Political economy dynamics

Phase II is characterized as early commercialization. It was affected by a number of factors simultaneously that did not collectively accelerate clean-energy technologies but laid a solid foundation for it, as depicted in Figure 1. This was a time in which natural gas, oil, and coal prices reduced nearly 50% compared to the previous two decades, which led concomitantly to only moderate policy interest

Niche Markets

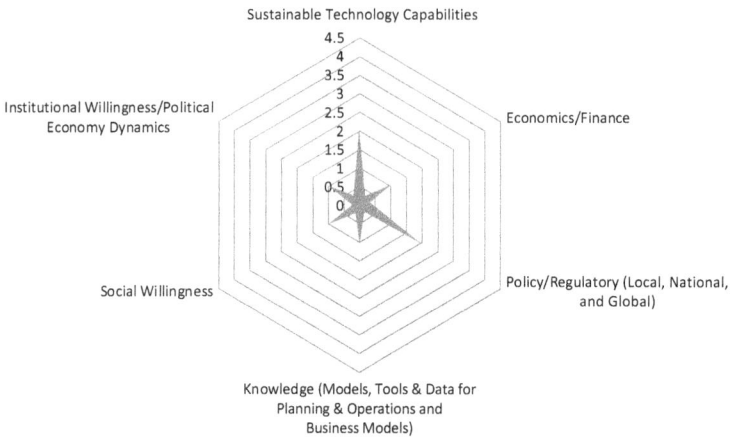

Figure 1. A changing set of factors for Phase II.

in renewables in most locations, with the exception of continued support in California, and the moderate expansion of renewable portfolio standards (RPSs) to 10 states, with early support in Europe. At the U.S. federal level, the Energy Policy Act of 1992 opened up a suite of policies to support renewable energies, energy efficiency, alternative fuels, electric vehicles, appliance standards, and multiple other provisions to reduce foreign oil dependence and support energy innovation. The Energy Policy Act of 1992 bill followed the Global Change Research Act of 1990 that established a U.S. research program on climate change. While this suite of policies began to open doors for clean-energy technologies, it was not until 1998 – when the Kyoto Climate Conference was held and the Kyoto Protocol called for all nations to reduce greenhouse gas emissions – that policy ambitions began to rise around the world. But, the Kyoto Protocol was not ratified in the U.S., and thus failed as a global agreement, and the world continued to negotiate for a global climate agreement for nearly 20 years, until the Paris Agreement in 2015.

During this phase, technical champions continued to work on efficiency and scaling improvements for solar energy and overall cost improvements for wind energy, but progress was relatively slow. There was policy-induced interest in alternative fuels, particularly ethanol for gasoline blending, as well as initial interest in "the hydrogen

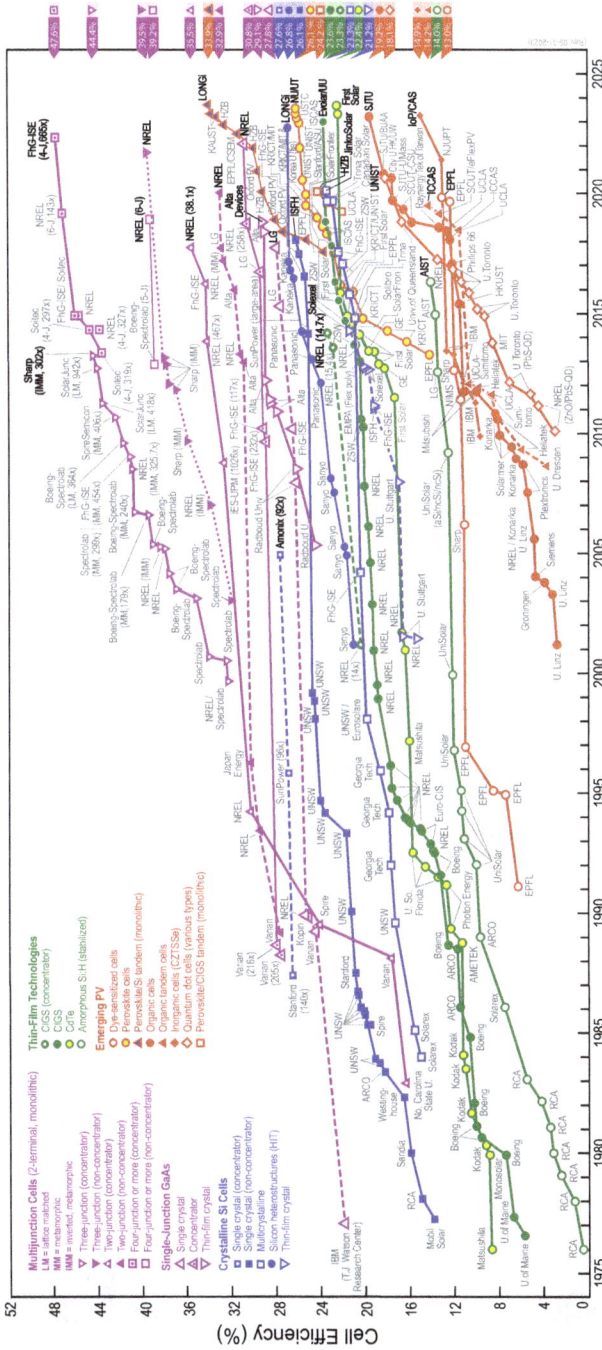

Figure 2. *Best Research Cell Efficiencies for Solar PV. Figure shows efficiencies for lab-scale PV cells for multiple technologies, both 1 sun* and under concentration, since 1976.*

Note: *1 sun refers to the testing conditions which are defined for solar cell efficiency tests to reproduce solar radiation at the surface of the earth. The conditions are standard illumination at AM1.5, or 1 kW/m².

Source: Reprinted with permission from NREL.

economy," which led to some efforts but did not lead to substantial commercial markets. NREL has produced a chart of solar cell efficiencies since 1976, as shown in Figure 2. One can see a relative stagnation of technologies over this time period (~1990–2000).

On the cost front, however, photovoltaic (PV) solar energy was going through substantial cost reductions – nearly a factor of 2x reduction from 1990 to 2000 alone. Studies analyzing these cost reductions began to be published, and focused on the scalability of renewable energy technologies. Classically, this is called endogenous learning, a concept that arose from the reflections of colleagues at the International Institute of Applied Systems Analysis (IIASA), who were then evaluating technological changes in energy technologies.[4] They introduced the concept of endogenous learning in order to better capture the cost reductions they were observing as well as market penetrations of different energy technologies. This led to a fairly substantial amount of new knowledge being created, elucidating multiple factors that affected learning curves, including both economies of scale and technological progress.[5] Similar to Moore's law that expressed the increase in transistors and computing power, learning curves were used to extrapolate the potential cost reductions for renewable energy technologies. That simplified approach created more enthusiasm for rapidly expanding the scale of the industry and doubling down on R&D efforts. What was missing in those early formulations, however, was an understanding of the numerous factors which affected cost competitiveness, including global material availability and pricing, and financial cost for "emerging" technologies that investors were not familiar with. These other factors came in to play early on as well as impacts of a few global dynamics (particularly for the steel industry for wind towers), and for photovoltaics where it was shown that some supply chain constraints affected global learning rates. However, the global effects were short-lived and the powerful effect of economies of scale for photovoltaics and engineering advancements for wind continued to support substantial cost reductions for these decades. Cost reductions of approximately 20% for every doubling of the market continued to build enthusiasm and interest to expand the market. Manufactures and project developers were creatively leading businesses to ensure profitability. These were not without exemplary failures, including many with large oil and gas parents, such as ARCO Solar, Mobil Solar, BP Solar International, and independents such as Energy

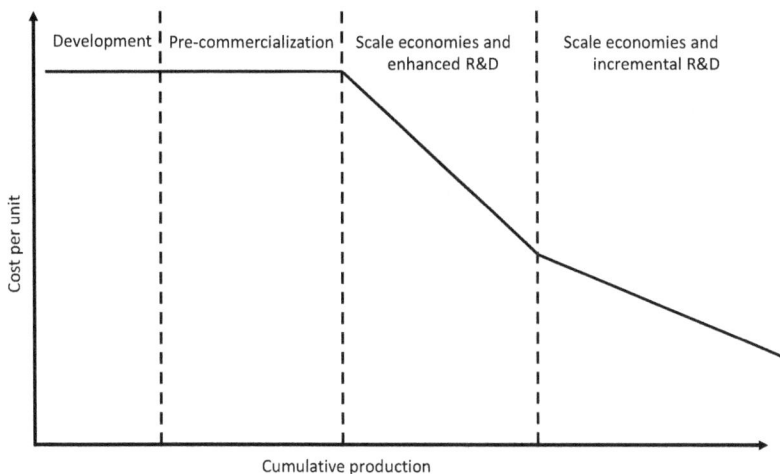

Figure 3. Technology learning curve showing the four stages of classic technology learning as knowledge and economies of scale lower production costs.

Conversion Devices (ECD) and Evergreen Solar. Failures such as these are not unique to the solar industry. There are hundreds if not thousands of examples across the IT space and other hard tech spaces, particularly those that are capital intensive and have long return cycles.

A conceptual learning curve is shown in Figure 3 in which four stages of technology costs are broken down into functions of technology maturity and market size. From a cost perspective, technologies begin at a relatively high cost because they are still under development and have not been proven or manufactured at scale. This persists for quite some time, potentially many years to decades, depending upon the speed and scale of the market. To be clear, this has been applied to nearly all "hard" technologies. That is, the learning and economies of scale do not apply to software platforms that, of course, leverage a huge installation base of computing, the Internet, smart phones, etc. and propogate to new markets readily. Significant learning is seen in the third stage where the technology effectively reaches a tipping point in the market, moving from early adopters to mass markets. During this stage, at least two factors are at play: first, economies of scale and/or material supply as well as cost per unit output from manufacturing scale; second, incremental and potentially substantial improvements in performance based upon rapidly scaling research and development related to market growth. The last stage is a slowdown in the overall learning speed along with cost reductions as a function of cumulative deployment as the technology matures. In this stage, depending upon the application, new emerging technologies may increase in competitiveness, particular if they can scale and become cost-effective.

This book will present further reflections on technological learning in Phase IV, from the perspective of 2022, so that one can understand nearly 40 years of technological progress and what that has offered and will offer for the future of economies based on renewable energy.

As for wind energy, there is ample literature on the modeling and physics of wind turbines at a fundamental level.[6] Simply stated, wind speed increases with height from the earth surface, reflecting surface boundary layer effects. Wind power increases as a function of wind speed cubed, or to the third power. That implies that taller wind turbines can effectively harness more power for a given site. Larger wind turbines, that is, those with longer blades, have the ability to extract more energy from their "rotor swept area." Increasing the height of the turbines introduces further engineering challenges relating to the weight, wind forces, and extreme conditions, such as turbulence, hurricane force winds, and ice storms. Regretfully, during these early commercialization decades, many wind facilities experienced failures and catastrophic incidences that gave wind a relatively unreliable reputation. But, engineers caught on quickly. Wind turbine technology rapidly grew from 500-kW 50-m height turbines to 7.5-MW turbines at 140 m of height. The scaling of height and rotor diameter along with advances in gearbox designs, controllers, and power electronics brought substantial improvements to reliability and economics for wind power generation.

The following was succinctly stated by my colleagues in a recent paper on the "Grand Challenges in the Science of Wind Energy"[7]:

> Three fundamental drivers have reduced the cost of wind energy to date: increased hub height, power rating, and rotor diameter. These can be understood using the fundamental equation for wind turbine energy capture:

$$P = \frac{1}{2}\rho C_p \left| AV^3,$$

where P is the instantaneous power produced, ρ is the air density, C_p is the power coefficient (or overall machine aero-dynamic, mechanical, electrical performance measure), A is the swept area of the rotor, and V is the freestream air velocity. The design of the machine impacts access to higher velocities, V, as well as performance, C_p, and the attainable area, A. Increasing the hub height reduces the influence of the surface friction, allowing wind turbines to operate in higher-quality resource regimes where wind velocities, V, are higher, with a compounding effect on power production. Larger generator capacity coupled with power electronics – which enable variable speed operation – provides more power produced per machine installed at a given location (assuming a constant C_p). More power per turbine allows fewer turbine installations and lower balance-of-system costs and fewer moving parts (for a given level of power capacity), thereby enhancing reliability.

To get a sense of the growth of wind turbines over time, Figure 4 gives a sense of scale – note the relative size of the Airbus 380 aircraft in comparison.[8]

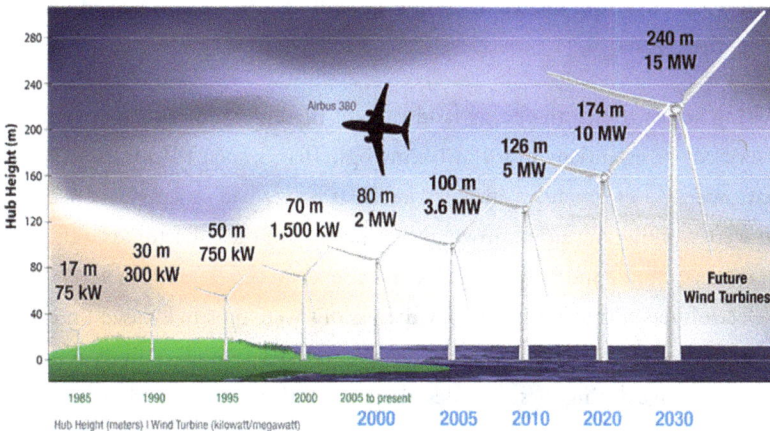

Figure 4.　Evolution of wind turbine size and height.
Source: Reprinted with permission from NREL.

Figure 5 shows the global cumulative installed capacity (in gigawatts) for wind energy and estimated levelized cost of energy (LCOE) for the U.S. interior region in cents per kilowatt hour from 1980 to the present. The cost data reflect the improved engineering and increased heights and power extraction per delivered

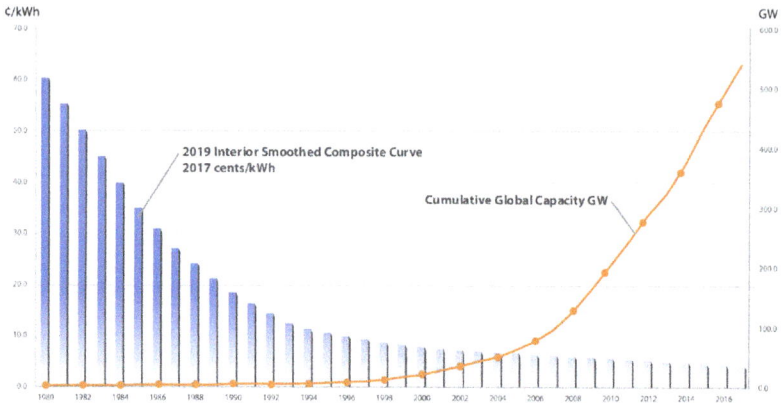

Figure 5. Cumulative installed capacity and LCOE for onshore wind energy.
Source: Reprinted with permission from Ref. [9]. Data from GWEC, IEA, IRENA, NREL.

unit of power. From the early 1980s through early 2000s, the cost reductions are more a function of technology progress as opposed to manufacturing economies of scale, as noted from the relatively slow growth in cumulative installed capacity during these years.

Looking across the clean-energy landscape of these decades, the approach that persevered was essentially siloed by technology. That is, solar PV advanced within a narrow market segment. Concentrating solar power had independent activities that were working toward improving the efficiency of the collectors as well as the storage medium. Wind power engineering also advanced. Hydrogen was focused on electrolyzers. Building technology discussions were mostly focused on energy efficiency improvements. But, because oil, gas, and coal were all relatively abundant and inexpensive during these decades, there was little if any discussion on electric vehicles, smart homes, or smart appliances.

Economically, renewables remained expensive compared to other clean-power options or clean-fuel options. Their scale was not significant enough to reduce overall costs. Wind costs declined relatively steadily during this phase, as noted in Figure 5 above. This was the result of relatively intense engineering improvements that allowed for taller and larger turbines which could withstand and extract the energy from higher wind speeds at greater heights above the ground. Markets were still quite niche, prompted by policy environments, particularly in Denmark and California.

Photovoltaics, on the other hand, made relatively little progress technologically or economically during this phase. Cell efficiencies, as shown in Figure 2, did not improve markedly, but costs continued "down the learning curve," lowering from ~$10/W to ~$5/W for crystalline PV modules between 1990 and 2000. This pricing on modules was not accompanied by concomitant price declines in the balance of systems (racking, mounting, inverters, permitting costs, and sales), which approximately doubled the installed price, inhibiting market expansion. For a sense of scale, global PV production capacity in 1990 was only about 200 MW and grew to ~1,000 MW (1 GW) by 2000 (compared to over 250 GW in 2020 and on track toward 1,000 GW/yr by ~2025).[10]

From a policy and regulatory perspective, Phase II included instrumental expansion of policies to support clean-energy technologies, with the introduction of tax credit provisions in the U.S. Those provisions, the production tax credit (PTC) and the investment tax credit (ITC), have offered the most substantial federal policy support since their first introduction. The initial approaches for these tax credits limited the lifetime of the tax credit to 10 years. This proved to be helpful to get some initial projects financed and built. The PTC offered a tax credit for each MWhr of production from a qualifying facility. The ITC offered a tax credit to the capital investment of a qualifying facility. Nominally, the PTC applied to wind systems and the ITC was used to support solar power plant investments. Authorized in 1992, the policies had little material impact on the scale of the market for many years due to a combination of factors, such as relatively moderate RPS targets, immature markets for power purchase agreements, and perceived technology risk in the capital markets. As other factors, such as increasing state RPS targets, began to expand, the overall scale of deployment grew. Onshore wind installations had an initial uptick just after the PTC was authorized in 1982, growing to nearly 500 MW installed in 1985, but then dropped off and remained below 500 MW/yr until 1999, when nearly 600 MW were installed. Then, wind installations went through three boom-and-bust cycles as the PTC was in place, then expired, and was then passed again. Installations in good years were 1,500–2,500 MW between 2001 and 2005, but as low as 200 MW when the PTC expired.[11]

The ITC was also nominally impactful on solar energy in the early 1990s and 2000s as costs and other policy drivers (e.g., RPSs) were not sufficient enough to attract developers and investors. The initial 30% tax credit was only authorized for two years – barely enough time for developers to secure permits and complete a project. This introduced a stop–start dynamic into the U.S. marketplace through the 1990s that had a significant impact on manufacturers, project developers, and financiers who were fighting for capital and institutional support to build renewable energy businesses in a strongly competitive environment where capital and talent were also being sought out in the oil and gas industries. It was not until 2008 when an extension was negotiated that had reasonable durability to provide confidence in the markets.

At the state level, policies that were established in Phase I in California continued, with some new policy introduction, with some expansions, but little innovation. The expansion of state RPSs in the U.S. spread to AZ, CT, IA, NJ, NV, MA, ME, MN, PA, WI, and TX by 2000. Each state had a slightly different approach and, by today's perspective, low or moderate targets (e.g., 10–30% by 2020).[12] Federal programs were focused on energy research and development. Some state policy experimentation, mostly oriented to small systems, and relatively small budget allocations at the state level were adopted. As of March 1997, 39 U.S. states had some form of renewable energy, energy efficiency, or alternative fuel policy or incentive program. Of the 11 states that did not, two had property tax exemptions still in effect for structures built prior to 1985 but not for new buildings, three supported alternative fuel programs that did not explicitly mention ethanol, and five had legislation specifically indicating the need to reduce reliance on foreign sources of energy by supporting the development of renewable energy resources. The most common incentive type was some form of loan or grant for energy conservation, renewable energy applications, innovative energy projects, and/or energy research. Examples of state approaches included loans or grants for small systems, property tax incentives, personal income tax credits, corporate tax credits, sales tax exemptions, and industry recruitment incentives.

In Europe, in December 1990, the first national feed-in tariff legislation (FIT) was adopted by Germany's Electricity Feed-in Law (Stromeinspeisungsgesetz, or

StrEG).[13] As of January 1 1991, utilities in Germany were required by law to buy electricity from non-utility RE generators at a fixed percentage of the retail electricity price. The StrEG included a purchase obligation for this electricity and the percentage ranged from 65% to 90% depending on the technology type and the project size. A project size cap of 5 MW was also imposed on hydropower, landfill gas, sewage gas, and biomass facilities. Denmark and Spain followed suit with similar provisions in 1992 and 1997, respectively.

The FIT approach was further improved when certain municipal utilities in Germany began offering FIT prices based on the actual costs of RE generation (a model pioneered by the cities of Hammelburg, Freising, and Aachen), primarily to encourage solar PV. This cost-based framework enabled efficiently run projects to be profitably operated and this design feature continues to be identified as one of the most attractive models today. This approach was in contrast to an avoided cost or "value-based" approach to tariff calculation, or one in which the prices were tied to the prevailing retail price. Avoided cost could be based on the cost of power that would not be needed, such as coal- or natural-gas-fueled power. Value, on the other hand, can be calculated via multiple approaches. One approach utilizes the effective cost of delivered power to the location of generation (e.g., a house with PV). In many billing approaches, this includes the avoided generation cost, transmission costs, and distribution costs. Multiple other approaches have also been proposed, but many of these have not gained significant traction since the transmission and distribution systems were recognized as necessary to provide service whenever the house could not generate sufficient energy from solar PV.[14]

The European Commission adopted one of the first quantitative target policies for renewable energy. In 1997, it established a 12% goal by 2010 for renewable energy production.[15] Next, the Renewable Energy Sources Act (Erneuerbare Energien Gesetz, EEG) was adopted by the German Parliament in April 2000. This was the result of a campaign by Hans-Josef Fell, who was a member of the German Parliament, a leader of the Green Party in Germany, and vice president of EuroSolar. This legislation signaled a number of important developments: (1) FIT prices were decoupled from electricity prices at the national level; (2) in contrast to previous FIT policies, which focused primarily on fostering non-utility generation, utilities were allowed to participate; (3) RE sources were granted priority access to the grid; (4) FIT payments for wind power were differentiated by the quality of the resources at different locations; and (5) FIT prices became methodologically based on the costs of generation for all technology types.

Somewhat in contrast to Germany's approach, Spain's RD 661/2007 introduced an innovative "sliding premium" option for FIT-based renewable projects. This design offered a variable FIT payment or premium above the spot market price, which effectively ensured that project revenues would remain within a range sufficient to ensure profitability. Policymakers could use this option to increase the market integration of RE sources because electricity was sold directly on the spot market and received an additional FIT payment. In April 2008, the Netherlands adopted a similar framework, where a sliding premium covered the difference between the prevailing spot market price and the guaranteed FIT price.[16]

From a knowledge perspective, renewable energy penetration in power markets was quite low (<5%) and not problematic. There were few if any studies looking at increased penetration of renewables and the potential implications of that for power system operations. In fact, technical reviews published in the early 2000s timeframe predominantly focused on singular technology performance. A good example of this is the article written by Thomas Ackerman and his colleagues on the status of wind power.[17] This article provides an excellent historical perspective as well as cataloging the engineering advances and early market penetrations in multiple markets around the world. From a technical integration standpoint, scientists and engineers focused on the power quality output of singular technologies, that is, a wind turbine or PV inverter, as opposed to, for example, a wind farm or a portfolio of renewable energy assets connected to a regional- or country-level grid. Broader system-level knowledge advanced greatly in Phase III, led by a few, including Ackerman, into a new science of systems-level grid integration of variable renewables. This book will touch on that and much more in later sections.

From a societal perspective, the FIT approach in Europe and state-level efforts in the U.S. were beginning to get noticed by a segment that was predisposed to clean-energy solutions and entrepreneurs who were relatively early movers in the

clean-energy development segment. Overall interest across the U.S. or in Europe was relatively lacking. Development agencies and those focused on rural energy access, rural energy healthcare, and water production seemed to be the groups most focused on renewable energy solutions. This was a product of the reality of the very limited grid supply and energy access in developing countries.

Institutionally, across the breadth of stakeholders – utilities, regulators, consumers, developers, and policy and regulatory decision-makers – there were pockets of interested leaders such as Fell in Germany, but little overall interest. Provisions that were put in place in U.S. states were perceived of as important experiments, but were not perceived of as a major market signal or as a major threat to existing business models, utility economics, or the economics of large energy incumbents.

However, these experiments laid solid groundwork for what was to come in Phase III. The dialogue and intention of the Kyoto Agreement and Kyoto Protocol raised the consciousness of and awareness toward climate change and the need for a transition of our energy system. This of course was not until the late 1990s, and it took multiple years for a few nations and their energy ecosystems to begin to respond. This was led by the development of a solar strategy in Japan and the expansion of the solar industry in Germany and across Europe. Wind energy also began to take off in Denmark and northern Europe. Real momentum was building.

Chapter 3
Phase III: *From Problem Child to Problem Solver* – The Era of Exponential Growth (2000–2020)

Key Themes

- From niche to competitive

- Market expansion and policy innovations

- Germany, China, U.S., and Ireland

  o Lessons on policy

- Technology scaling, economy of scale, expanding into main markets

  o Operational challenges from increasing penetration

  o Early operational technical solutions, approaches to integration

- Analytic modeling: Insights, ambitions, and implementation

- Myths and misconceptions of integration

  o Rapid learnings

  o Tools and "flexibility"

- Least cost options in many countries

- Hybridization

- And:

  o The promise of returns

  o Visions for electrification

o   Cross-sector coupling and power to X

o   A circular economy

Factors

A.   Sustainable technology capabilities

B.   Economics/Finance

C.   Policy/Regulatory (Local, national, and global)

D.   Knowledge (Models, tools & data for planning & operations and business models)

E.   Social willingness

F.   Institutional willingness/Political economy dynamics

Is it really possible to run a national power system on renewables? No one knew. Some had suggested that it was possible. Even back in 1999, a colleague, John Turner, "the Wiz," published a paper in the *Science* journal suggesting that 100 square miles of photovoltaics (PV) could provide all the energy the U.S. needed on a given day.[18] But, no one had really studied this. Power engineers certainly did

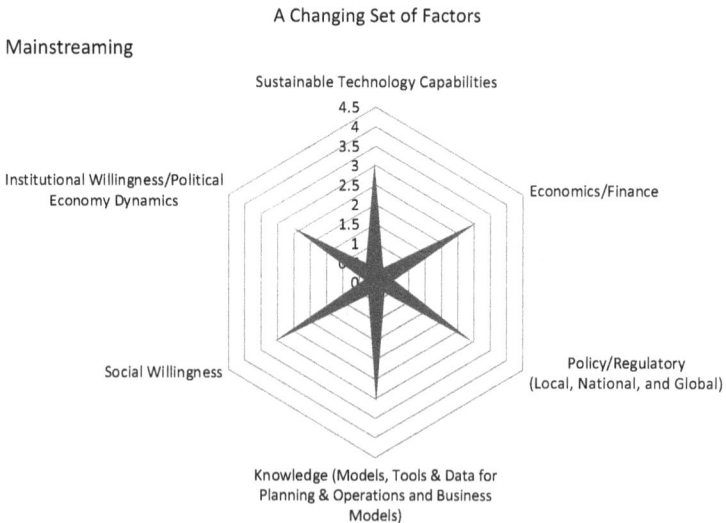

A Changing Set of Factors

Mainstreaming

Institutional Willingness/Political Economy Dynamics

Sustainable Technology Capabilities

4.5
4
3.5
3
2.5
2
1.5
1
0

Economics/Finance

Social Willingness

Policy/Regulatory (Local, National, and Global)

Knowledge (Models, Tools & Data for Planning & Operations and Business Models)

Figure 1. A changing set of factors for Phase III.

not believe it was possible. Models were not capable enough, yet. But questions like that motivated hundreds to advance their thinking, modeling, and technical developments to begin to find answers.

Phase III covers the most recent two decades of rapid expansion of renewable energy technologies, in combination with rapid expansion of IT technology, communications, computing, modeling and forecasting, and global collaborative peer learning. It was, as many might reflect, an era of explosive growth of not only renewable electricity generation technologies but also technology solutions that addressed demand-side management, energy efficiency, electrification of mobility, and the introduction of business model innovations that would affect both capital expenditures as well as the operating expenditures and cost effectiveness of energy solutions. It was also the era of fracking for natural gas, complex energy geopolitical dynamics, rapid expansion of social-media-enabled communications, and social movement influence. The scientific and technical literature, textbooks and courses, and available trainings were extensive. In this relatively concise book, this era is characterized by the six factors affecting the enabling environment and success of clean-energy technologies. In doing so, a significant amount of other work is referenced and left for the readers to digest on their own. Stylistically, Figure 1 captures the changing landscape of the six factors, showing significant changes from Phase II.

Statistics for this era are rather stunning. Wind installations grew by more than 100 times and solar installations grew by 700 times to more than 700 GW cumulatively by the end of 2020. In 2020, the capacity of global wind installed was 93 GW with a total installed base of 743 GW according to the Global Wind Energy Council, more than twice the total installed global nuclear power capacity. Annual PV installed capacity in 2020 was 138 GW, with a cumulative capacity of 773 GW. More poignantly, annual renewable power additions grew substantially; more than 80% of all (including natural gas, coal, and nuclear) power generation capacity installed in 2020 was renewable, with solar and wind accounting for 91% of new renewables, according to the International Renewable Energy Agency (IRENA).[19] By the end of 2020, the total global renewable generation capacity (including hydro, geothermal, biomass, solar, and wind) was 2.8 TW. Total worldwide power generation stood at 7 TW.

U.S. Solar Cost and Capacity History

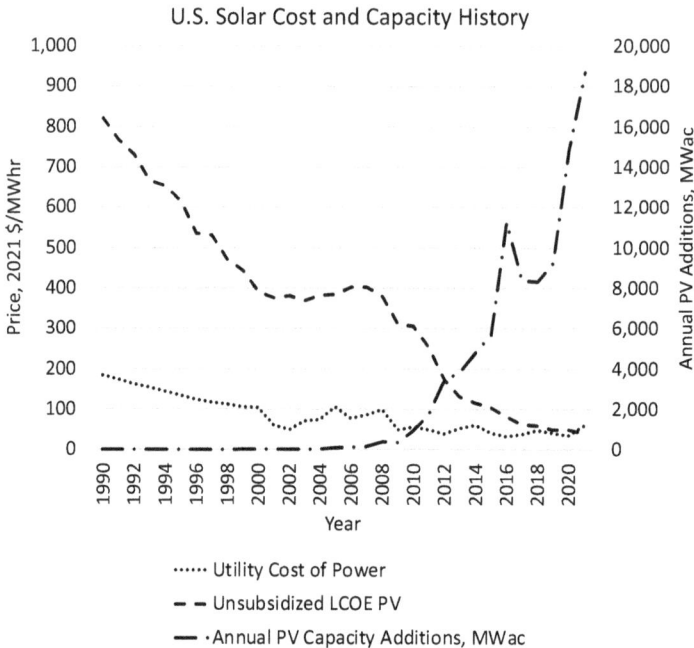

Figure 2.   Annual PV capacity additions in MWac, unsubsidized LCOE, and utility cost of power for U.S. utility-scale installations.
Source: Data from NREL.

In the U.S., PV installations took off exponentially, as shown in Figure 2, growing from below a GW to more than 13 GW annually by 2020. The reduction in annual installations experienced in 2018–2019 was due to temporary expiration of the ITC. The cost of power from PV dropped precipitously as well, from close to $400 MWhr in 2000 to ~$40/MWhr by 2020, excluding the tax credit provisions.

Capacity, however, is only one indicator. Many prefer to look at TWhrs of production, particularly owing to the fact that average solar capacity factors (CFs) are around 25% and wind capacity factors range between 35% and 55%. That means that every gigawatt of solar energy produces ~25% of the equivalent gigawatt of a 100% operational (capacity factor) power plant (which in fact does not exist in reality given maintenance schedules and other outages). Geothermal plants run above a 90% CF. Many existing nuclear plants, for example, have annual capacity factors above 90%; natural gas plants range from 20% to 85% and coal-generating stations range from 25% to 75% (average U.S. coal fleet capacity factor was 53% from 2012 to 2022).[20]

Further, from a technological perspective, this was also an era of expansive innovation in energy efficiency technologies, communication technologies, particularly cellular telephone communications and the Internet, and the use and value of data and data management. Some specific examples are noteworthy: innovation in lighting in particular had an incredible impact in developing countries. The transition from incandescent bulbs to compact fluorescent to LEDs allowed renewable-based home systems to transition from very expensive 100+Watt-level systems to micro and pico systems (e.g., 5–20 W) that could provide effective lighting as well as communications and charging for cell phones at increasingly affordable and attractive prices. Other innovations in developing countries included rapid expansion of cell-phone-based payment systems in combination with controls, remote monitoring, and, from a business perspective, innovations in business models which enabled significantly more affordability through incremental payments versus capital expenditures.[21] Interestingly, from a business perspective, the fact that consumers had a cell phone was a signal of credit worthiness as well as identification in some locations where both were hard to establish, difficult to come by, or not normal.

It must also be recognized that a significant technology revolution was taking place in many other sectors. For example, computing, the Internet, and cell phone and mobile telephony were undergoing massive transformations. Much has been written about each of these separate technology revolutions; thus, the interfaces between and among them – with energy technologies and in particular renewable energy technologies, smart meters, distributed controls, and computing – were beginning to be explored to enable the continued revolution between energy technologies and the Internet of Things. This is addressed in depth in Phase IV.

In developing countries, where energy access was a significant challenge, and persists today in many countries, this was also a time when non-profits, aid agencies, and experts recognized that measuring the number of systems installed or the number of kilowatts or megawatts deployed was in fact only an initial and insufficient indicator of progress. What became more important was the value of the system to the users, the ability to avoid theft, the ability to employ local installers and maintenance staff, and the ability to support not only education and communications but also small businesses and economic development. New thinking brought increased recognition that energy was a critical enabler of economic development rather than a consequence of economic development as part of a fundamental change of perspective. That is, neoclassical economic theory as it relates to energy use assumed that economic activity and the ability to pay enabled greater purchasing power for modern energy services. What became increasingly apparent during this Phase was the fact that access to modern energy was a critical enabler for economic development and prosperity. The availability of power to access high-quality evening education, communications, and economically productive activities such as sewing increased the productivity of chicken farming, water pumping, fuels, and mechanized machinery and improved agricultural productivity leading to increased economic productivity, creating a positive feedback loop in the economy.[22]

Looking for an anchor tenant who had a clear value proposition, such as a cellular telephone relay station or a small shop, both of which could then extend clean power for services such as cell phone charging or water pumping and purification, became a means of introducing clean-energy economic development, expanding from micro systems to mini grids, and offering pathways out of poverty. Here, the technology, business, and theory of progress perhaps form a glass half full perspective, as it remained by 2020, but there were still some 700 million people without access to modern energy, a reduction of only around 300 million in the preceding 20 years. This recognition led to the establishment of the sustainable development goals, or SDGs, in 2016, which was a relatively substantial advancement of global commitment to eliminate energy poverty and the related goals of ensuring access to water, education, food, and sustenance.

Economics

The most widely communicated observation of the economics of renewable energy in this phase is the significant (i.e., 90%) drop in the nominal cost of energy (in particular for wind and solar energy) over these two decades. Other technologies such as geothermal energy or large-scale hydro energy did not experience such major cost reductions. Nuclear power was reported to be increasing in cost due to multiple factors, including its long construction duration, and increased safety and reliability requirements.[23] Coal-fired power generation and natural gas systems became more efficient through engineering advances, but the cost economics for those were and remain heavily dependent upon fuel cost. Additionally, there was, and continues to be, an ongoing dialogue on whether or not to internalize externalities in project or technology costs.[a] There is a very large body of literature on the social cost of carbon and how that may or may not affect project-based economics.[24] For our purposes here, project-based economics are considered without taking into account the externalities. However, from an economy-wide perspective, as discussed in later sections, with increasing recognition of the cost of health and environmental impacts, and climate change including extreme weather events, the macroeconomic conversation also continues to evolve. This is more poignant going forward, but had been increasingly gaining attention during the latter part of Phase III.

To understand technology-based economic comparisons, the industry had utilized a singular metric for many years which was generally expressed as the levelized cost of electricity (LCOE) produced. It is important to flesh out the components of this indicator, starting with the simple levelized cost of electricity (sLCOE) as expressed in the following equation[b]:

sLCOE = {(overnight capital cost * capital recovery factor + fixed O&M cost)/(8,760 * capacity factor)} + (fuel cost * heat rate) + variable O&M cost,

where capital recovery factor is CRF = $\{i(1 + i)^{\wedge}n\} / \{[(1 + i)^{\wedge}n]-1\}$, i is the interest rate (or weighted cost of capital), and n is the financing period in years.

Capacity factor (CF) is a fraction that represents the percent of time in a year that the system will provide its nameplate capacity output. If a plant were to run

---

a Internalization of externalities is a classic term of art used in economics to reflect translating environmental impacts, which could be costs or benefits, into the economics of both private sector calculations and social good calculations. Examples might include the cost of impacts of pollution, implied loss of productivity, or increased morbidity and mortality in economic calculations of either the private sector or the social sector.

b For a more detailed calculation and details of power generation technologies, see atb.nrel.gov.

8,760 hours, its CF would equal 1. Typically, solar PV plants have CFs in the 20–25% range and wind farms in the 0.3–0.6 range. Many nuclear power generation facilities have CFs of 0.9, accounting for required scheduled maintenance.

Heat rate is the efficiency of burning a fuel into power. It is expressed in Btu/kWhr. The higher the heat rate, the lower the overall efficiencies. This is highly dependent on the system's technology. For example, natural gas combined-cycle turbines have heat rates in the 6,000–8,000 Btu/kWhr range, while gas steam turbines have heat rates of 10,000–12,000 Btu/kWhr. Coal steam turbines have heat rates of 9–11,000 Btu/kWhr, with supercritical coal plants being more efficient.[25]

Thus, while effective levelized cost of energy for wind power dropped nearly 90%, with the same for solar energy, evaluating the components of LCOE provides the critical insights that are important for our understanding of technological evolution. Most studies focus on the overnight capital cost element – this is the classic analysis of economies of scale and the learning discussed earlier, which affects the $/MW of the technical system. This, of course, is the key determining factor that has changed so dramatically over time.

Other factors, including cost of capital, have varied over this period of time, affecting the relative LCOE. Perhaps most interesting, for renewable technologies, the perceived risk in the capital markets has reduced over time. This has translated to a lower overall cost of capital as the technologies matured and as financial instruments became more robust in addressing the relative risks. In these two decades, innovation occurred in the capital and financial markets, which should not be ignored. In Europe and the U.S., this evolved from relatively simple capital structures of debt and equity to complex structures, which included tax equity investors, multiple tiers of debt, hedging instruments against which contracts were financed that included weather risk derivatives, performance-based provisions, and contract structures based on a combination of fixed and variable revenue streams. During this phase, there was a significant uptick in project-based finance, from a few billion per year to more than $500 billion per year. Much if not all of this came in structured financial arrangements for contract-based power purchase agreements.[26] There was and continues to be a conversation regarding how to expand capital access for the clean-energy transition, which is anticipated to cost many trillions of dollars per year in the coming decades. Here, discussions include those led by Joseph Stiglitz, Nobel Laureate economist, focusing on opening up long-term social capital structures, including sovereign wealth funds and pension funds, which have rarely allocated significant funds to project finance versus security-backed annuities. There is a deeper conversation as well regarding transitioning macroeconomic metrics to include indicators of environmental health, e.g., green GDP.[27] While those conversations continue to progress, there is no definitive pathway forward. A few sovereign wealth funds have aligned their investment portfolios with a clean transition. This in some sense reflects a growing consciousness in the asset management and capital markets regarding the responsibilities of fiduciaries, particularly those with long-duration funds. The 2000s also brought a palpable dialogue around environment, social, and governance (ESG) investing, starting from a small group of investors who initiated socially responsible investment as a core strategy back in the 1980s. Steve Schueth, who was the Vice President of Socially Responsible Investing for Calvert Funds in 1989, led a team that aligned investment strategies with core values such as clean-energy technologies and climate change. Having worked with Steve and his colleagues for more than a decade, it was remarkable to reflect on the growth of assets under management, positively correlated returns in environmentally sound technologies and companies, good governance practices, and social practices including taking care of employees and community engagement. Using their influence on board resolutions and in coordination with dialogues and actions with the Security and Exchange Commission (SEC) and others such as the United Nations Environment Program who have been working with central banks, there has been a significant increase in investments for clean energy as well as more broadly ESG-aligned assets during the 2010s. This was accompanied by strong voices calling for divesting from fossil fuel assets and the end of further fossil fuel investments, followed by multiple moves by university endowments and asset managers. However, the industry had not developed rigorous standards, and the ESG approaches face ongoing scrutiny.[28,29]

Deep commitments to environmental investment are rare. In 1995, while in Washington, D.C. working for the U.S. Department of Energy, I met Ian Simm, who had a strong interest in clean-energy technologies, energy for development, and asset management. He and I brainstormed over how to create a scalable, financially viable photovoltaic industry in South Africa during a trip there. That led to many months of business plan developments and seeking potential investors. Through those conversations, Ian was successful in raising funds from International Finance Corporations to begin Impax Asset Management, which was initially focused on photovoltaic market transformation. Ian never wavered. He looked for scalable asset management models with a sound thesis and good companies to invest in. The lure, or perhaps the myth of the lure, of venture capital or technology innovation investment did not attract him. There was no economy of scale in the business model. There were numerous success stories of unicorns and billionaires being created. Let's hope those continue as well. But, core investments through scalable models that can reach capital markets, pension funds, and sovereign wealth funds are truly needed to bring trillions of dollars to bear on clean-energy transition. Upon reflection in 2022, it is fascinating to note that it took until 2007 for Impax to grow its assets under management (AUM) to 1 billion U.K. pounds, but AUM grew 10-fold over the next 10 years and stands at nearly 40 billion today. Recognized twice by Queen Elizabeth II of the U.K. for its environmental leadership, Impax provides proof and hope that aligned investment will continue to scale to support energy transition over the following decades with the trillions of dollars needed.

From a technology capital cost perspective, most of the news was focused on the cost per kilowatt or per megawatt of a given technology, that is, dollars per kilowatt of a solar panel, or dollars per megawatt for a wind turbine or installed geothermal facility. This particular cost component saw enormous reductions over a period of time, as a result of learning by doing and learning through innovation elements as mentioned earlier. Significant innovation went into manufacturing scale and processes, uniformity, and effective utilization of manufacturing capital. Reductions in the use of materials, particularly in the solar PV industry in which cells became thinner and thinner, and thus less and less costly, led to additional cost reductions. Operation and maintenance costs, i.e., both the fixed and variable costs, also reduced overtime, reflecting the relative maturity and competition in the services industries and engineering.

What is not reflected in this simple LCOE approach are explicit details which can be important, such as project finance cost, depreciation and tax treatments, and construction costs, but which can vary dramatically when one is building, for example, a large nuclear plant over 10 years versus a gigawatt-scale solar plant that might be built within 1–2 years. Grid connection costs can also be included as one-time cost expenditure, or they can also be expressed as part of the fixed or variable operation and maintenance costs.[30]

Where does competitive economics come into play? Here, the industry evolved to using LCOE as an effective metric for comparison across renewables, natural gas, coal, and nuclear power, including all the subvariants and technology options. While this was useful in understanding "delivery of electrons" to the grid (at the point of generation or interconnection to the grid), it was less than useful in providing insight to well-risk-adjusted investments across the full system as it focused only on generation and effectively positioned each generation technology in competition with any other, as opposed to having sophisticated approaches to hybrid systems, full system dynamics and costs, or potential synergies among technologies. For example, comparative economics over the life of the projects involved in many cases sophisticated modeling of comparative energy costs. Project economics are a function of technology and development costs, as well as cost of capital, market structures (revenue), and financial hedging. Using natural gas as an example, gas prices fluctuated quite dramatically over multiple decades as shown in Figure 3. Projecting natural gas pricing with any certainty at any given time would be challenging. Thankfully, the use of more sophisticated models and modeling to account for

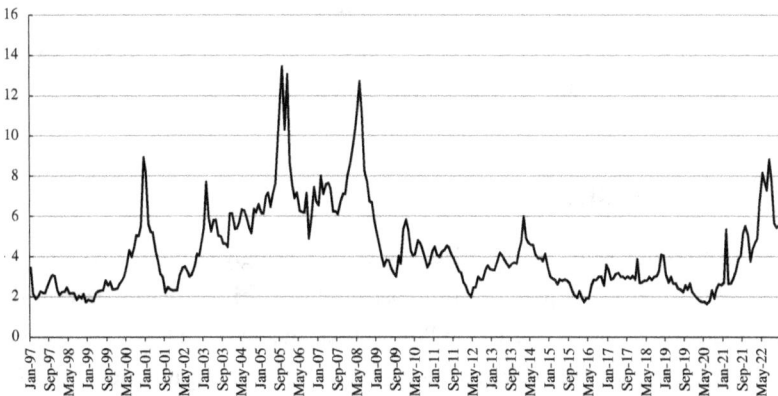

Figure 3. Henry Hub natural gas prices from 1997 to mid-2022 (dollars per million Btu).
Source: Data from EIA.

uncertainty is available for financial analysts as well as energy modelers. However, in the 1990s, those tools were not widely used or as sophisticated. Projecting price fluctuations at that point was not standard practice. Similarly, relief from the high gas prices of the early 2000s, the rapid expansion of the resource base and production, and the dropping of prices resulting from the fracking boom were also not foreseen.

The natural gas ecosystem started undergoing fundamental changes in the early 2000s when George Mitchell used hydraulic fracturing ("fracking") in a commercial well. His commercialization of this technology, which was created through early research at U.S. Department of Energy, set off a "titanic" change in the outlook of natural gas in the U.S. Resources estimates exploded from 40 TCF to 464 TCF, leveraging new technologies for resource characterization and production.[31]

The advent and mass commercialization of horizontal drilling in combination with fracking set off a massive movement to acquire mineral rights (in the U.S. where they are privatized) and investment for production. Globally, countries were reassessing their own resource potential and also watching and learning from the U.S. where developers were expanding well development in close proximity to schools and neighborhoods, which spurred a new movement against close-proximity oil/gas development and a flurry of state regulatory actions. It also spurred innovation in the well development industry with creative companies using electric (as opposed to noisy diesel) equipment and sound barriers.[32]

It further spurred a rethinking of "competition vs cooperation" for natural gas and renewables relative to the operational synergies for operating power systems, and also in comparison to coal for reducing GHGs.[33]

Policy Dynamics

The early 2000s were characterized by a fascinatingly complex dynamic of government policy and public–private partnerships. This was epitomized in the solar industry. Following the Kyoto Protocol, Japan announced the formation of the New Energy and Industrial Technology Development Organization (NEDO) and expanded production (the Sunshine and New Sunshine Program) and market mechanisms following the early experiments of feed-in tariffs. As noted previously,

this market policy approach was initiated in California and then expanded in Germany. In Japan, the combination of feed-in tariffs and government support for PV manufacturing led to relatively rapid growth of a PV manufacturing industry. This proved viable for a short while, until the Chinese program expanded at an unprecedented rate and competition drove the German producers out of business.[34]

China's growth in renewable power generation as well as production of solar and wind technologies was also significant in this era. Spurred by a combination of demand-side and supply-side policies, availability of low-cost capital, and coordination and alignment with expertise from the Chinese Academy of Sciences, the Chinese solar industry expanded dramatically. Chinese production of PV skyrocketed from virtually zero in 2000 to 2.5 GW in 2007, to 50 GW in 2016. This dwarfed production in Germany, Japan, and the U.S. in 2016.

The global policy landscape to support renewables evolved substantially as well. From only a few countries supporting renewables in the early 2000s, nearly 150 countries had some sort of supporting policy by the end of 2020. These policies leveraged early experiments and approaches, including FITs and renewable portfolio standards. The policies also expanded into fuel standards such as those included in the energy policy acts of the U.S. and Brazil. Fiscally, many countries also adapted tax policy considerations or other fiscal incentives to support renewable power as well as the use of renewables in heat and fuels.[35] Toward the end of the 2010s, with increasing focus on climate-related policies, a few of the national climate policies included explicit provisions for renewable energy in obtaining decarbonization goals. This was a leading element from the U.K. and the European Commission. The initial proliferation of policies to support electric vehicles emerged, and funding to support electric vehicle charging infrastructure became substantial enough to provide some confidence in the visibility to the expansion of low-carbon transportation in alignment with clean electrification. Additionally, by the end of 2020, global interest in hydrogen resolidified, based on some leading analyses which indicated that net-zero power systems combining electrons-to-hydrogen for clean chemicals and fuels for heating, power, and industrial and building applications would be essential to obtain decarbonization goals, retain a robust energy economy, support jobs, and support industries around the world.[36,37] Ambitions for green hydrogen capacity were relatively modest, but in the many tens of gigawatt scale. This was seen both as a driver of innovation and an opportunity to exploit low-cost renewable power generation for industrial export, for example, in Chile and Germany.

In the U.S., in 2008, an eight-year extension to the PTC/ITC was granted which provided some policy assurance to developers and investors. This assurance, the changing costs for solar energy in particular, which dropped from $400/MWhr in 2000 to $200/MWhr by 2012, and state policies expanding to more than 30 states with some type of RPS or renewable target, many with a "set-aside" target for solar, provided collective support for rapid growth. Annual solar installations only amounted to about 10 MW in the U.S. in 2000 and grew to over 10 GW/yr in 2016 and nearly 20 GW/yr in 2020.

During this time, innovations were also occurring in the commercial contracting and financial engineering of projects. As discussed earlier, PURPA initiated the concept of qualifying facilities being able to contract for delivery of power. This began with developer–utility contracts until about 2005. At that time, a few large companies, mostly in IT, that were focused on clean energy and climate mitigation as a principle began to advance corporate PPAs. These are essentially bilateral contracts between a developer and a corporate energy buyer, with clarification of the attributions of the power (e.g., renewable). This market was quite small initially, with only a capacity of 650 MW in total between 2008 and 2012. It then began to take off as PPAs for renewables could offer price certainty and in most cases very attractive prices as opposed to the fluctuating costs of buying power in the market. By 2015, more than 2 GW of renewables was contracted under the corporate PPA structure. Multiple types of synthetic PPAs – contract for differences, options, and commodity hedges – were being created and contracted. Stated in what might be overly simplistic terms, the contract for differences allows the price to be pegged to the going market rate with a differential adder or multiplier. Options contracts follow traditional structures from the financial sector in which the purchaser may buy a set amount of clean energy and hedge the pricing with put-and-call options to reduce the price risk. Other structures began to emerge such as bifurcated projected output where a portion of the energy is sold under a virtual PPA at a hub price and the remainder is sold behind the meter or partially merchant. These structures continue to evolve, reflecting the creativity of financial managers and deal brokers. Financing renewable energy projects with corporate PPAs has been more challenging than financing projects with standard utility PPAs due to the often lower credit ratings of corporates, the more frequent fluctuations in power demand, collateral allocation, and other issues. Every surveyed bank seeks a first lien on specific project collateral in loan documentation underpinning corporate PPAs.

During this phase there was ongoing attention for the extension of the PTC and ITC in the U.S. Expansion of corporate goals and desires for renewable energy procurements (and attractive prices for power as well as tax appetites) led to a relatively rapid expansion of the tax credit markets via financially engineered contracts and the ability to bring tax credit capital into project structures. Further financial engineering creativity was bringing in resource risk insurance and hedges as well as other risk mitigation structures that reduced the perceived risk for investors and developers. Toward the latter part of this phase, a few leading corporate purchasers, including those from the IT sector who were seeing massive growth in their power demands and who had corporate commitments to sustainability and science-based targets, began discussing the challenges associated with 24 × 7 × 365 procurement of delivered clean electrons versus purchasing RECs or through PPA instruments. Their motivation was to drive further development of the markets but also have directly attributable carbon mitigation impacts. This corporate interest, in combination with evolving market structures and increased revenue for dispatchable power in times of scarcity, prompted multiple developers to create hybrid system approaches, integrate storage, and develop concepts and projects for 24 × 7 dispatchable renewable-based power.[38] This in fact became part of the overall auction approach in India and in the United Arab Emirates where they have procured specifically renewable power for between 8 pm and 8 am to offset the production of a significant amount of solar photovoltaics during the day.[39]

Similarly, in Europe, the policy landscape continued to evolve to reflect the growing knowledge of operating power systems with more renewable energy technologies, cost reductions, and the growing desire to address decarbonization goals. At this time, Europe promulgated renewable energy targets for each of the EU member countries.[40] This was the principal driver of continued development of renewable energy projects within the European Union. However, each country was able to approach the RPS at its own discretion. For example, some countries like Germany and Belgium were strong advocates of feed-in tariff approaches and distributed energy resources. This contrasted with, for example, Denmark which was strongly advancing large-scale wind and offshore wind turbines, reflecting the resource base and attractive economics for that country. In contrast to Germany and Denmark, among others, which dispersed the supplemental costs of the policy to rate payers, other countries such as Spain developed a feed-in tariff structure for projects which alternatively had a financial support mechanism which was initially tied to the government central budget. The financial collapse of the late 2000s led to abrupt restructuring of the financing terms and to multiple lawsuits and eventual restructuring of the financial incentives that were covered through the electricity sector versus the central government.[41]

Knowledge: Many of the technological advances in Phase III were brought about by a combination of economies of scale for manufacturing, advanced sensing manufacturing controls, engineering knowledge, and the ability to scale installations. For utilities, many of which had not operated a system with more than a few percent of generation from variable renewables, the technological advances represented the initial set of challenges for integrating variable supply as opposed to just dealing with variable but relatively consistent demand. (That is, daily load profiles were relatively well known and understood in most power districts. These changed over time throughout the seasons, which was also well understood.) What was new, however, was learning how to integrate generation that changed with cloud cover or wind patterns. This, in hindsight, produced early reactions along the lines of "this is a problem that we should not have to deal with" or these are "unmanageable" resources. However, leading utilities and system operators were willing to share their knowledge and experiences. They were required to learn how to integrate more renewable generation technologies by the market requirements. In the U.S., for example, in California, renewable maximum penetration grew from approximately 15% in 2012 to nearly 65% in 2018, with an explosive growth in solar energy from just a few percent to nearly 40% of the peak generation. Similar growth was experienced in Texas and the Southwest Power Pool, but dominated by wind energy. In Ireland, renewable electricity shares grew from approximately 8% of annual electricity in 2006 to greater than 40% in 2020.

Instantaneous generation percentages by renewables were allowed to expand to 65% through 2020 and 70% from 2021 onward.[42]

Other examples include Portugal running for multiple days on 100% renewable energy in 2016. These incidences offer glimpses of the potential of renewable generation for annual reliable zero-carbon power, but also offer a stark reminder of the challenges of operating a very-high-percentage renewable grid. Other countries including Brazil, Uruguay, Denmark, and Norway run power grids that are predominantly low-carbon and renewable, utilizing a significant amount of hydropower resources in combination with some wind, solar, and biopower. In some of those countries, Brazil in particular, recent droughts have brought about concerns over the stability of the power system and have spurred interest in other renewable resources including solar, onshore and offshore wind, and even more recently green hydrogen. Brazil is also a leading country using biomass for power generation. This is in coordination with their clean-fuel program for sugarcane-based ethanol. This is an example of early integrated systems thinking for resource optimization and use across multiple sectors.

It was not until the mid-2000s that models were advanced enough for us to begin to ask the question: Could 20% of the annual electricity be supplied by wind power in the U.S.? This question inspired the 20% Wind Energy by 2030 integration study, the first of a series of studies evaluating the potential of renewable energy technologies for the U.S.[43] Published in 2008, the study was enabled by the creation of a new modeling approach that took into account the geospatial and temporal characteristics of renewable energy technologies and combined those with geospatial and temporal economics. The brainchild of Walter Short at NREL, what was then called WinDS for wind deployment system, the model was a fundamentally new approach toward power system modeling as explained by his (along with Nate Blair and Donna Heimiller) article in *Solar Today* in 2003.[44] By using high spatial and temporal resolution datasets for wind and solar energy, Walter and the NREL team were able to create economic supply curves at a relatively high resolution for the continental U.S. He combined those economic supply curves with statistical treatment of the variability of renewables by creating representative time windows in which supply had to meet demand as well as reserve margins for reliability considerations. Wrapped into an economic linear optimization framework, the model was able to compare the economics of developing solar energy in region X versus wind energy in region Y, including the transmission from regions to the demand center which required power in that regional transmission area. On top of that was the option to explore increasing interregional transmission or intraregional transmission. This capability began to open up the opportunity to ask the following questions: Is large-scale transmission more cost-effective than developing localized resources? How much long-distance transmission might the U.S. need to develop for renewables? Technological representation was also important because in each of the supply curve areas, one had the option to develop renewables, natural gas, coal, nuclear, and storage, or bring in power to meet demand through transmission. Thus, the development of more approaches to represent the future performance of various technologies as well as forecasting of fuel prices was needed for more robust insights.

While the U.S. 20% Wind Study led to more questions about the capability of evaluating potential futures to give confidence to energy system planners, policymakers, R&D program managers, and researchers, it was not until 2012 that the first ever national report on renewable electricity futures was published. Commissioned by, and the brain child of, Sam Baldwin at the U.S. Department of Energy, this large body of work was the first rigorous evaluation of a national power system based upon renewable energy generation. In that study, scenarios were evaluated in which renewables provided between 30% and 90% of the total electricity generation required for the U.S. A core scenario of 80% renewables (all technologies – hydropower, geothermal, biomass, wind, solar, etc.) was compared and contrasted to other approaches for decarbonization of the power system or the economy based upon more economy-wide models that did not have the technical details required for power system reliability analysis.[45]

There were multiple key elements in this groundbreaking effort. First, it included experts from over 35 organizations in multiple working groups, working together over three years. It was the first deep dive into evaluating the future cost and performance of renewable energy technologies in a rigorous approach that had been mapped out and thought of in terms of learning by doing and learning through innovation. Further, it was a time when analysts had to rigorously define the differences between technical potential and economically developable potential.

Here, technical potential is the calculation of the full resource potential of a given technology to produce power or provide fuels, for example, the number of photons that hit the U.S. any given day converted to GW potential or terawatt hours per year. From a renewables perspective, these numbers are enormous. Accounting for areas that are not suitable for solar development, the technical potential of solar energy alone is in the 190 TW range, e.g., 190,000 GW (compared to the full 173,000 TW of sunshine the world receives continuously). Wind technical potential is about 10,000 GW of energy, geothermal about 35 GW of energy, biopower about 100 GW of energy, and hydropower about 200 GW of energy. Recall that the total installed production capacity in the U.S. is around 1,200 GW. From these technical potential numbers, analysts then had to take into account sensitive land areas, high slope areas, national parks, wilderness areas, etc.[46] For offshore wind, assessments were broken down by distance from shore and depth of ocean. They could then calculate the economically developable potential at high spatial and temporal resolutions, e.g., 2 km × 2 km for the continental U.S., with 7 years of data down to a 5-minute resolution.

Utilizing an enhanced version of the WinDS model, now called the Regional Energy Deployment System model or ReEDS,[47] in combination with higher-fidelity power system operational models which could look at hourly operations of the power system for 8,760 hours in a year, the teams were able to evaluate much more rigorously the potential trade-offs among renewable technologies, the impact of transmission or its constraint, and requirements for new storage. Model outputs included economic metrics such as total system costs, fuel costs, and wholesale power prices, and other metrics such as carbon dioxide emissions, NO_x, SO_x, and water usage. The latter developed into a whole new stream of work on the energy–water nexus, which is a subset of a larger framework of integrated energy, economic, environmental modeling that will be described later.

Results from the Renewable Electricity Futures Study indicated that the U.S. could have reliable power with at least 80% if not 90% renewable electricity generation. Of that 80% or 90%, approximately 45–50% would be derived from variable renewable sources, that is, wind and solar energy. It was important to distinguish these given the dialogue of the time and understanding of the potential challenges of integrating variable renewable resources into the operations of a power system. The study, given the economics at that time, indicated that wind energy would be the predominant new renewable resource with solar energy being a close second. New transmission would be an important component, but if that were constrained, there were sufficient localized resources to also be able to meet the 80% scenario goal. If transmission were constrained, lower cost resources that would be further away from load centers would not be developed at the same scale. Total system cost would increase compared to more localized resource development. Storage was relatively expensive in 2010. At that time, the U.S. had approximately 20 GW of installed storage, predominantly pumped hydro systems. In the 80% case, the economics indicated that ~100–150 GW of storage would be developed by 2050. This would be developed in order to meet the reliability requirements, as they were characterized. Interestingly, in 2022, the economics among various technologies, including wind energy, solar energy, and storage, have all changed from 2010. More recent studies confirm that 80–90% renewable or zero-emission power systems can be reached economically and reliably. The technology mix, however, is dramatically different than that evaluated in 2010. We will look at these results in more detail in Phase IV.

The Renewable Electricity Futures Study was the start of a campaign led by non-profits and other organizations to advocate for federal, state, and local authorities to conduct similar analyses and promulgate policy rules and regulations to advance renewable portfolio standards or other approaches for advancing decarbonization in the U.S. For example, the results of the Renewable Electricity Futures Study, in combination with rapidly decreasing costs and advocacy efforts, led multiple states to increase their RPS targets, along with multiple regulators approving significant expansion of renewable energy within the integrated resource planning processes due to their cost attractiveness.

This was also a time of significant progress in energy systems modeling and scenario analysis. The academic literature was growing in terms of articulating methods and approaches. In combination with increased availability of high-resolution datasets, the ability to manage large datasets with faster and larger computers led to similar studies being conducted for Australia, China, Denmark, and broadly across Europe.[48] A collection of those analyses at this stage identified a few key insights which were transformational at this point in time, but interestingly have remained consistent for the next decade. These include the importance of transmission expansion as an option to bring low-cost renewables from remote areas to demand centers and smooth out variability among different locations of renewable generators. Next, it included the concepts of flexibility and enabling power markets. Here, flexibility was introduced to highlight the importance of being able to manage demand more than in the past, when demand was essentially taken as granted and it was the job of the utilities to meet demand at any given time. This also meant flexibly operating generating assets which may have been initially designed for "baseload" operations. This led to the introduction of a new lexicon for power systems and generator characterization. That is to say, traditionally, generators were characterized as baseload, mid-merit, or peaking. These constructs were used within the institutional and governance infrastructure which built the electric power system in the U.S. and most countries around the world. Under a regulated system structure, the utilities' principal purpose was to provide reliable, affordable power at a reasonable rate of return. That meant that the utility could propose a set of assets and operating principles to the regulator, which if approved were built and used, and in which the utility was remunerated by a set of regulated rates. This framework began to change as some markets moved from regulated to restructured, that is, competitive wholesale markets. There is a separate body of literature on competitive wholesale electricity markets to which we refer the reader.[49,50] For our purposes, one can reflect that competitive markets are one particular construct for the governance of power markets, as is a regulated market structure. If the principles and goals are set out clearly for decarbonization, economic efficiency, reliability, and resilience, one can envision effective mechanisms within multiple governance structures. The debate of regulation versus competitive markets is left to others as the evolution of renewable energy technologies is further explored, and how it enables the achievement of those principles and goals independent of the market structure.

What was becoming apparent was that all the generation and preferably as much load as possible would benefit from becoming increasingly flexible or increasingly manageable and called upon by the grid operator. This utilized the early work of the Danish energy system in which the stakeholders were making changes in order to accommodate increasing amounts of wind. Their approach proved foundational for power system operators to both require and utilize flexible output from coal generators, which were initially designed for constant output. They asked a simple question: Could our coal plant outputs be reduced and then increased as we needed them to in order to maintain reliability of the power system to accommodate the changing generation of our wind generators? The answer was yes. In fact, the Danes found that their coal fleet could ramp 95% of its rated output and still operate reasonably. It did have economic impacts (reduced energy sales) and they also found that some of the equipment would be stressed and would degrade at rates that were different from their original design criteria. This concerned the engineers and financiers who had anticipated that the plants would operate at given levels for many years, as this flexible generation would both decrease the energy output, and thus energy sales, and shorten the life of the plant.[51] At this stage, utilities and system operators were also becoming increasingly comfortable integrating relatively low penetrations of renewables (less than 50%) into their system and had not yet considered scenarios within which renewables would provide a majority of the power or that they would seek to deliver 24 × 7 × 365 clean or net-zero power.[52,53] Those were later developments.

Also in 2012, a best practices paper on integrating variable renewables was published.[54] In that, it was articulated that there were five fundamental approaches for energy leaders to undertake. Succinctly, the five key actions are:

1. Lead public engagement, particularly for new transmission.
2. Coordinate and integrate planning.
3. Develop rules for market evolution that enable system flexibility.
4. Expand access to diverse resources and geographic footprint of operations.
5. Improve system operations.

Substantial public engagement was required to overcome the myths and perceptions, particularly related to transmission expansion, which is highly valuable for integrating diverse geographic resources to enable greater predictability and ease system operations through spatial aggregation. This, of course, is related to two other fundamental principles, which are to lead the coordination of integrated planning, and to expand access to diverse resources in geographic footprints. The third key action was to lead the development to creatively engage in market design and regulatory changes. It was already recognized at this point in time that least cost economic dispatch would allow zero marginal cost resources to provide most power in energy markets. However, in many parts of the world, dispatch was highly regulated and operations were planned and managed with potentially large time steps such as week-long blocks versus best practices of five-minute intervals. Shorter operational time steps allow for greater economic efficiency, as well as the development of rules to incentivize system flexibility. Lastly, it was important for leaders to champion the modernization of system operations, such as integrating in state-of-the-art solar and wind forecasting for power systems dispatch.

Stakeholder engagement was seen as critical to allow for effective representation of different perspectives in the planning process. An inclusive planning process which employed best practices to define conditions for power sector scenarios in long-term planning was an emerging best practice. That is, the future scenarios should reflect the desires and concerns of the local stakeholders. Some examples are as follows: Can my system operate on 100% renewables? If so, at what cost? Would a system that included fossil fuels with carbon capture and storage be viable? Where would the CO_2 be stored? Is that safe? What about hydrogen or renewable natural gas? Is biomass power acceptable? What is the potential for rooftop or other distributed solar? Where and how many EV charging stations do we need by 2050? What will the power requirements look like and what are the trade-offs for different system configurations? By answering questions such as these, scenario analysis allows the stakeholders to explore issues of interest and concern. By using state-of-the-art power system models and techniques, system operators who ground their decisions on the physics-based operations and planning models can have confidence that their system will operate reliably and economically with increased penetration of renewable energy technologies. The technical confidence was a prerequisite to increasing political and stakeholder ambitions for higher renewable energy penetration in power systems.

Following a very similar policy and energy planning trajectory that began during the oil crisis of the 1970s, Denmark set forth a plan to reduce dependence on imported oil. The initial steps of the plan introduced coal power as the alternative. Nuclear power was proposed but rejected as an alternative in the 1980s. During this time, there was growing interest in wind power and finally by the mid-1990s the technology had developed sufficiently to become increasingly cost competitive at sufficient scale. The Danish policy landscape included numerous approaches to support domestic wind development, including feed-in tariffs and then a transition to renewable portfolio standards. It also allowed for co-op development of wind generation plants which allowed families to invest in and benefit from wind power development generation. This, along with the evolving national and international policy landscape and willingness of the population and the system operators to make deeper commitments to renewable energy generation, allowed the Danes to be at the forefront of the integration of best practices. By the early 2000s, Denmark had developed approximately 2.5 GW of wind power, providing ~15% of their annual electricity generation. With a power system that is interconnected with that of Germany, Sweden, Norway, and the Netherlands, the Danes were early pioneers in integrating wind power, introducing flexibility and showing the value of interconnections.[55] Building on work in Denmark on wind integration and the flexibility of their coal power generation units, with insights from the REF Study and international practices, it became more widely recognized that the increase in penetration of renewable energy in power systems required system operators and market designs to enable greater system flexibility. Flexibility here meant that all assets in the power system, including demand and supply, had to be more dynamically responsive to market signals and system requirements. This led to three principal approaches. First, establishing markets of shorter time duration led to economic efficiencies while assuring or improving overall system reliability. Second, all assets should be remunerated. Third, bigger balancing areas would bring substantial benefits, if there were coordinated dispatch with neighboring balancing areas, coordinated unit commitment, or more fully consolidated operations.

It was at this time that early experiments showed that renewables could provide highly accurate generator responses to signals from a system operator. The groundbreaking proof of this was provided in California in 2016 where a 300-MW solar plant could "provide essential reliability services related to different forms of active and reactive power controls, including plant participation in (auto generation control) AGC, primary frequency control, ramp rate control, and voltage regulation," with extremely high accuracy. Results showed that PV (or other renewables that had high-speed power-electronics-based interfaces with the grid) could provide "regulation accuracy," that is, the ability to provide the requested power output at 4-second intervals (the time requested by the system operator) at 24–30 points

better than fast gas turbines.[56] This soon became an emerging best practice for utilities and was introduced into grid codes and performance requirements.[57] However, in the U.S. and many other countries, it was contrary to the PPA contracts and financial incentives of the production tax credits which were oriented toward maximum energy production and not toward being a "good grid citizen." Other lessons that were being learned included the value of day-ahead and real-time forecasting.[58] This inspired a whole new industry to convert weather forecasting into energy forecasting and integrate it into control rooms and system operations.

This was also the time when substantial skepticism arose regarding the potential role of renewables in power systems. The following questions were being asked: Can renewables provide reliable power? Does each individual renewable plant (wind or solar) need "back up"? (This was assumed in many energy and integrated assessment models which erroneously increased costs. New physics-based modeling offered new approaches to characterize the system more accurately that could then be incorporated into less complex models and reduced the conservative cost escalation.) Is storage required on a 1:1 basis to make sure renewables are "dispatchable"? Is it not expensive to integrate renewables into the grid?

These perspectives around the role and impact of renewables and power system operations were pervasive as people were questioning the prior approaches toward systems design, reliability requirements, and operations. Such perspectives were also pervasive in the energy modeling community which at the economic level had to come up with creative solutions to stylize the relative representation of renewables in power systems. These were early days. Models such as EIA's National Energy Modeling System (NEMS), while a flagship for the U.S., had particular representations reflecting concerns over requirements for dispatchable power and constraints on renewable power growth, somewhat reflecting on supply chains or economics. Classic energy economic models such as those used and championed by MIT are called computational general equilibrium (CGE) models. CGEs such as these have economic representations of the capital and financial flows within an energy economy. They do not however have the representation of physical systems. It is a difficult challenge to represent, for example, the subtleties of power systems in those models. Collaborative efforts where one can couple more detailed models with the CGE models have been undertaken and shown to be valuable for enhanced and improved understanding of the details of some sectors. This is also the case in models for other countries and perhaps more poignantly the integrated assessment models which are the foundation of the IPCC. Most of the models in the early 2000s, because of their global nature, had poor representation of the details of power systems. Because they had to work at the global scale, they worked on a relatively coarse spatial scale such as national or even regional. Resource information was drawn from the global datasets at 1° resolutions, which are enormous when one considers the spatial requirements for doing high-resolution power sector modeling. For those not familiar with the calculus, the 1° resolution of a global model equates to something on the order of 100 km. They also worked on 100-year time frames, typically out to 2100. That meant that the timestep in the calculation was at best one year but more likely every five years. Of course, power systems have to operate in real time and thus it was a difficult challenge to represent that level of resolution over such a long-term energy/climate model. State-of-the-art renewable resource assessment and representation in energy models is at a resolution of 1–2 km, if not higher. Also, for the highest fidelity temporal modeling, typically five-minute datasets are used, which contrast to average annual or perhaps multi-seasonal representations of resource capability. Recognizing these limitations and the advancements that were happening in our understanding of the ability to model smart, detailed complex systems, a number of collaborative efforts were undertaken. The community of energy economic modelers embraced these challenges with open arms and were coordinated and spearheaded by Professor John Weyant of Stanford University, who with his colleagues has led energy modeling workshops for many decades.[59] Over the next decade, many improvements were made not only in the country-specific models but also in the global integrated assessment models. Some of the implications of these improvements can be seen in the scenario results that were published as early as AR3, compared to the 1.5°C special report,[60] and most recently in AR6.[61] Much more detailed analytics at the country level have also been advanced, such as Princeton's NetZero Americas,[62] and multiple efforts by colleagues at NREL on pathways to decarbonized power systems.[63]

Then, in 2017, a number of NREL colleagues presented their observations at an Institute of Electrical and Electronics Engineers (IEEE, which has over 400,000 members in 60 countries) Power and Engineering Society (PES) meeting that set forth the next chapter of understanding on how to integrate over 50% renewable penetration into power grids. Their observations were as follows:

Fact one: The grid can handle more renewable generation than previously thought.

Fact two: Geographic diversity and resource diversity provide additional reliability to the system.

Fact three: Wind and solar forecasting provides significant value.

Fact four: Our electric power markets were not originally designed for variable renewables, but they can be adapted.

Fact five: Modern power electronics are creating new sources of essential reliability services.

Each of these observations prompted deeper experimentation to prove that grids with a higher percentage of renewables could in fact be managed cost-effectively and reliably. Innovations were coming from California, Texas, Australia, Ireland, and many other locations. Notably, the renewable generation percentage in Denmark had grown to 74%, and many other locations were routinely handling 20–50% variable renewables instantaneously. There were numerous operating examples and studies which pointed to the benefit of greater geographic and resource diversity, adding to reliability. This was proved in a number of efforts that showed the smoothing effects of power generation across larger geographic areas and the flexibility of import and export of power through integrated transmission interconnections. Further, with increased resolution, both temporally and spatially, of forecasting data and models, it became apparent that system operators could forecast a day ahead and, with even better accuracy, improve hourly and sub-hourly production of their variable generation assets. This allowed them to operate the system with less stress, greater predictability, and lower costs.[53]

Knowledge and experience had been evolving from around the globe for the evolution of power markets with increasing shares of variable generation to more accurately represent the dynamics of real-time power generation and interactive loads. There was, however, a very active debate on market design and institutional restructuring. In one camp, there were continued arguments for vertical integration to evolve the power system within the integrated resource planning process. In that process, the regulated utility brings forward an integrated resource plan (IRP) to the regulator. In that plan, the utility lays out its rational and economic analysis for the development of new generation and transmission, as well as other assets as needed. In many instances, IRPs showed that additional renewable generation capacity at the wholesale level was indeed the least cost pathway and was approved by the regulator. These plans were brought forward even from the most conservative utilities given the cost reductions that renewables had experienced, increased knowledge and capabilities for operating systems, and stakeholder interests. Other market structures were also evolving. There was another camp of power system economists and analysts who were oriented toward fundamental restructuring of the industry, unbundling integrated utility into separate and competitive generation assets, system operators, and independent distribution companies. At the structural level, this approach proceeded in numerous locations in the U.S., the U.K., and elsewhere. The actual market rules, however, varied widely within these deregulated configurations.[64] Some of the leading energy and power economists had argued that all effective services could be represented via an energy-only market. In effect, the energy price would include "shadow" prices for other attributes that were required to operate the system. Such attributes could be required in grid codes and performance specifications. Others argued that it was more efficient and transparent to unbundle energy prices from other attributes. This included the introduction and evolution of capacity payments, transmission and distribution payments, and other "ancillary" services such as downward and upward ramping.[65] Further, as the markets were increasingly seen as a mechanism to enable flexibility to dynamically manage supply and demand, rules were being developed for demand response, in which large demand customers could be called on to reduce their demand during peak times and be paid for doing so. This also spurred innovations in demand aggregation. This concept involves a company connecting tens, hundreds, and even thousands or tens of thousands of devices via an integration platform, normally through IT-enabled communications, in order to provide effective services of the bulk power system. For example, one might consider thousands of controllable hot water heaters, air conditioners, or heat pumps, in which either the heater or the compressor is turned off for a few seconds or minutes. The end consumer experiences little to no change in their expected quality of service. That is, the water is still hot when called for or the fan continues to operate over a cooled heat exchanger and continues to provide air cooling for a few minutes while the compressor reduces the electric load. These types of innovations led to ~ 10GW of controllable peak load, realized either through load shifting or load reduction, in the 2000s in the PJM service territory. This avoided the need for more than $10 billion of capital investment and saved the consumers tremendous amounts of money while also enhancing the ability of the system operators to effectively manage the power system. Those customers that participated in this voluntary market received payment from PJM for the services they provided, amounting to nearly $500 M in 2010.

A comprehensive framework began to emerge to assess the various approaches for power system transformation based on integration of higher levels of renewables. Within that, there were six key components: system operations, services from the variable renewables, load management, flexible generation that is traditionally viewed as "baseload" generation, investment in transmission and enlargement of balancing areas, and storage technologies.[66] For example, in the system operation category, elements included improvement of renewable electricity forecasting, advancements in market design and system operations toward sub-hourly scheduling and dispatch, the evaluation, remuneration, and use of "flexibility reserves," and when politically and economically viable, expanding the footprint of balancing areas and/or considering energy imbalance markets or other joint system operations over larger geographic regions. The last option complemented the concept of increasing access to transmission as was previously mentioned. What was new, however, was the concept of deriving services from variable renewables. Here, technology innovation was unleashing the capability of inverter-based resources to provide voltage support, frequency support, auto generation control including "downward" reserves, and the ability to ramp upward and dispatch renewables. Previously, these were unthought-of capabilities. As already noted, these were exemplified in the groundbreaking demonstrations in California a few years earlier, but it took time to communicate that to the broader audience and also for them to have confidence that such capabilities could exist within their system as opposed to someone else's. Further, there was only a small cohort of system operators that had promulgated market conditions at the cutting edge of integrating renewables and implementing such innovations. Others, a majority, were still in the earlier stages of market and operational changes to accommodate greater amounts of renewables.

Storage, which has been used in power systems for a century or more in the form of pumped hydropower generation dams, was reemerging as a technically interesting area with the potential for integrating more battery storage and chemical storage, e.g., hydrogen or other chemicals. At this stage, most of the analyses and evaluations were focused on relatively short-cycle battery support for integrating renewables to complement solar generation profiles.[67] However, it had been known for decades already that concentrating solar power could be coupled with high-temperature oils or salt thermal storage which could be used to drive a steam turbine to provide dispatchable power.[c] Compressed air energy storage (CAES) was also well known, but restricted geographically. With the reemergence of storage as a concept, new ideas came forward including simple mechanical systems, such as the raising and lowering of concrete blocks and the use of the rail cars and railroad tracks to literally move a rail car up a hill to store it and allow it to return to a lower elevation and create power while doing so – a simple conversion of gravity, mass, and velocity to power.[68]

At the end of Phase III, scientists undertook a new evaluation of the potential for storage technologies for the U.S. power system. In this study, called the Storage Futures Study, I chaired a technical review committee which provided advice and guidance through a series of reports that were produced over approximately three years.[69] The study was motivated by a new grand challenge to reduce the cost of storage options to support decarbonization of the power grid, but also in recognition of the fact that storage cost, and in particular lithium-ion battery cost, was declining rapidly. This was predominantly driven by economies of scale from the light-duty electric vehicle market that was spinning off into power systems applications. The

---

c   The U.S. was one of the earliest developers of concentrating solar power (CSP) in 1984 with an initial capacity of 14 MW in Daggett, California. Fresnel lens/oil-based systems were built in the 1980s, following incentives in California and the PURPA act of 1978. Facilities with capacities of 354 MW were eventually built. Concentrating solar towers with 10 hr of molten salt storage have also been developed. In 2022, more than 1.3 GW of concentrating solar power capacity was operational in the U.S. and 6.8 GW worldwide.

value propositions being demonstrated in the power market spurred yet further innovation in alternative storage technologies including alternative battery chemistries. The study sought to answer a few fundamental questions:

- How might storage cost and performance change over time?

- What is the role of diurnal energy storage in the power sector, absent drivers or policies that increase renewable energy shares?

- How much diurnal grid storage might be economically deployed in the U.S., both at the utility scale and distribution scale?

- What factors might drive that deployment?

- How might increased levels of diurnal storage impact grid operations?

Relying on the increasingly sophisticated power system models that were available and the increasingly mature approaches to technology cost and performance projections, the study team evaluated hundreds of configurations and scenarios with a few different technology options. Evaluating trajectories toward low-carbon or zero-carbon power grids offered insights into the trade-offs between storage, generation, demand response, and transmission. Additional insights were derived relative to the costs and benefits of distributed storage and generation on the low-voltage grids or "behind the meter" versus bulk-scale storage and generation connected to the high-voltage grid. Leveraging what was anticipated to be the rapid expansion of battery markets and concomitant cost reductions for lithium-ion batteries, the team evaluated storage technologies and short-term applications, e.g., two to four hour applications, to the need for long-term or long-duration energy storage. Figure 4 shows the relatively recent significant drop in Li-ion batteries, an estimate of the continued cost reductions over the next decade, and an estimate of the growth in the market size broken down by use in EVs and other segments.[68] Note that there are the many different potential futures, and while the details differ, there is strong agreement on continued growth in battery markets and continuing cost reductions. Other battery chemistries would have to compete with the pricing and performance of Lithium-ion batteries, a few of which are

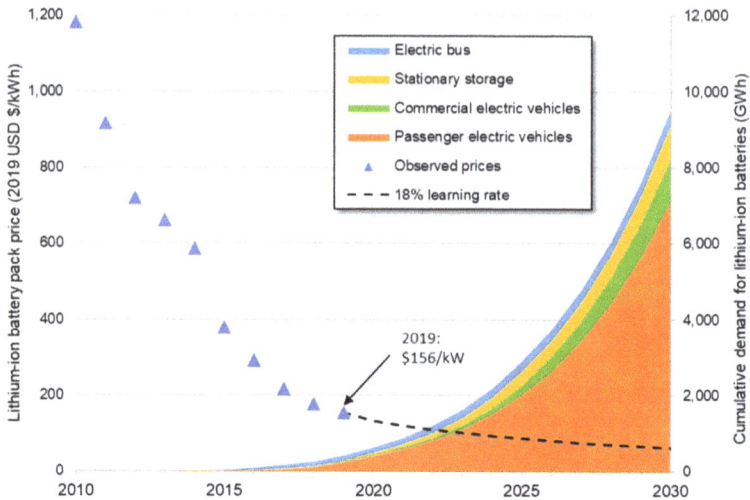

Figure 4.   *Lithium-ion battery pack price history and one projection going out to 2030 plotted along with anticipated growth in the demand for lithium-ion batteries in gigawatt hours (GWh) which is broken out by electric vehicles and stationary storage. Notice the use of a learning rate to forecast the battery cost reductions.*
Source: Reprinted with permission from NREL.

emerging but not at any significant scale yet. They may scale commercially in Phase IV and beyond.

Because storage is such a hot topic today, as will be discussed in detail later, including the potential role of hydrogen and other chemicals as long-duration energy storage, it is important to understand the application space for different storage technologies. Classically, this is mapped as a function of duration and energy as shown in Figure 5. In it, one can see multiple different technological approaches spanning compressed air energy storage to chemicals to mechanical to batteries as well as super-capacitors and other options.

The cost competitiveness of the different storage technologies will be a function of its application in the marketplace. For example, hydrogen or other power-to-gas options such as renewable natural gas could effectively be the energy source for peaking plants as well as long-run combustion turbines in times of scarcity for a power system that is predominantly based on variable renewables. Batteries are

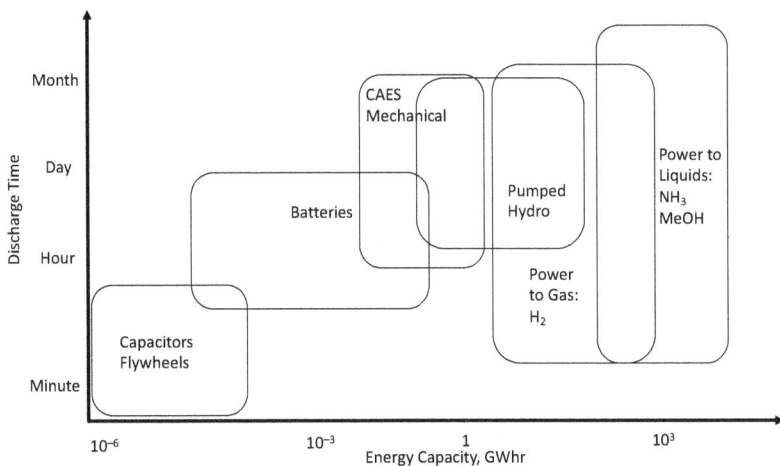

Figure 5.   Mapping of energy storage technologies as a function of the discharge time and energy capacity.

ideally fit for short-duration outages up to perhaps a day depending upon their costs. As of the early 2020s, batteries are economic for multi-hour applications and hybrid systems spanning from 2 hr to 8 hr. If the cost is significantly reduced, their application may be expanded to perhaps a day or two.

Long-duration energy storage has been found to be increasingly important for very highly renewable-based power systems; that is, for systems in which more than 90% total electricity is provided by all renewables (dispatchable and variable), with variable renewables being greater than 50% in order to provide reliability at the seasonal and multi-day levels.[70,71] The focus on these technologies continues strongly in Phase IV.

Key findings from the Storage Futures Study[68]:

- Storage is poised for rapid growth.

- Recent storage cost reductions are projected to continue, with lithium-ion batteries (LIBs) continuing to lead in market share for some time.

- The ability of storage to provide firm capacity is a primary driver of cost-competitive deployment.

- Storage is not the only flexibility option, but its declining costs have changed when it is deployed versus other options.

- Storage and PV complement each other.

- Cost reductions and the value of backup power increase the adoption of building-level storage.

- Storage durations will likely increase as deployments increase.

- Seasonal storage technologies become especially important for 100% clean-energy systems.

The largest integrated complex system that operates in real time was always lightheartedly referred to as the power systems around the world. While perhaps disputable, it was increasingly recognized that the grid itself, along with the flow of electrons and all of the physics details underneath it, was in fact an integrating function for multiple technologies. The architecture of the power grid includes multiple different types of generators, such as traditional fossil fuel plants, nuclear power plants, hydropower plants, wind farms, solar farms, and today increasingly batteries that are connected either directly or in combination with another technology. That is only on the generation side. On the demand side, end uses include heating and cooling (air-conditioning is one of the fastest growing loads worldwide), electrified mobility (EVs, as well as electric rail and electric aircraft), cooking appliances, water pumping and desalination, industrial processes, and industrial heating. All of these end use devices or processes are simultaneously connected to the real-time operating power system, which delivers power within very tight tolerances of voltage and frequency. The details of this power system are well covered in engineering textbooks. Recognizing the integrating function of a power system, along with continued advances in hybrid system design and operations, particularly for micro and mini grid systems in developing countries as well as for situations where reliable, islandable power was required, spurred further creativity regarding integrated hybrid configurations.[d]

---

d   Here, a hybrid system refers to an explicit design which combines multiple technologies, for example, solar plus diesel plus batteries, or solar plus wind plus batteries, or solar plus micro or pico hydro systems. Islandable refers to the ability to operate when disconnected (islanded) from the main power grid.

Small hybrid mini grid systems, of approximately a few kilowatts to perhaps tens of kilowatts, were initially developed for remote communities. In fact, interestingly, there was an alignment of the global sustainability agenda, which is predominantly viewed as the climate change agenda today, and the modern expansion of hybrid systems. Based on discussions at the Earth Summit in 1992, the U.S., working with others, launched what was then called the Village Power Program. The program was focused on hybrid systems for remote communities for provision of energy access to enhance economic development and reduce poverty. It morphed into what became the Global Village Energy Partnership, involving the World Bank, NREL, and many others. Early experiences uncovered a number of key insights. In the earliest systems, for example, the one developed in Mexico in 1982 in a remote community in Oaxaca, all of the system engineering design and installation costs were covered. It was essentially a gift to the local community. It turned out that lack of organizational structure, ownership, management, metering, and payment for services led to failure. This reflects similar experiences and prior learnings from the solar home system activities in which foreign assistance agencies and foundations focused on the number of systems that could be installed, but did not understand the human and institutional elements which were critical to the long-term success and ongoing use of those systems.[72]

Mini and micro hybrid systems were seen as quite attractive. This was typically in contrast to traditional energy and light sources or perhaps kerosene lanterns or candles, or even diesel generators. While most of those approaches in the 1990s or early 2000s were used for evening light, they were extremely expensive or dangerous. The danger side included horrific stories of spilled kerosene lamps creating fires or kerosene being transported in reused drink bottles, predominantly Coca-Cola bottles in many countries and communities across rural Africa. This inadvertently and sadly led to multiple incidents of poisoning of children.

Hybrid system engineering, governance, operations, and maintenance continued to evolve with expanded operations in more than 10,000 island communities around the world. Many of these communities were being supported by diesel generators providing power only for a few hours a day at very expensive prices and requiring constant delivery of fuel. Many of the locations were well endowed with both wind and solar energy, and hybrid configuration systems could deliver power at a third of the price and for many more hours of the day. Of course, the challenge was to find the capital required to support the engineering installation. In many locations, these needs were overcome by government programs that leveraged foreign assistance, multilateral development bank programs, and a small set of risk-taking entrepreneurs. Due to the fact that there was an existing business that provided power for the local community, the ownership, governance, and payment structures were more easily addressed in these island communities.

Interest in hybrid configurations continued to expand commercially, having found a compelling value proposition for the telecommunications industry for powering remote towers by solar energy and batteries, sometimes with diesel backup generators. Internationally, foreign assistance agencies began to develop programs for rural health clinics and rural schools. The military began to have an interest in forward operating bases and other remote locations. Software programs, such as that developed by colleagues at NREL called Hybrid Optimization of Multiple Energy Resources (HOMER), were created to allow relatively quick system optimization analysis. First developed in 1993 by Peter Lilienthal, HOMER became the de facto program for hybrid system design worldwide. Produced commercially in the early 2000s, HOMER has supported more than 1 million system designs and configurations and continues to be the go-to training software for young engineers and entrepreneurs.[73]

While HOMER focused on rural and remote communities, optimization and evaluation software continued to evolve. Colleagues at NREL then developed a tool called renewable energy optimization, or ReOpt.[74] This tool was able to evaluate either remote or grid-connected systems for a given site, given location, or load, and then a set of costs and performance criteria for various technologies. With a unique approach that allowed it to calculate the trade space among various configurations, it could also be run in batch mode over hundreds to tens of thousands of sites and could produce an overall portfolio assessment of opportunities. ReOpt was used early on to support the evaluation of tens of thousands of cell phone towers in the U.S., hundreds of industrial processing facilities, and other applications. Capabilities expanded to include integrated storage and to evaluate resiliency of islandable micro grids which are of interest not only to the military but also for critical facilities such as police stations and hospitals.

In the mid-2010s, large-scale integrated hybrid configurations were just being introduced in the thought leadership literature. These new approaches to integrated systems included tightly coupling generation from multiple sources, for example, nuclear, low-carbon fossil, and renewables in combination with multiple potential output streams.[75] Tight coupling is different from the integrated configuration of the power system as described earlier. Here, the multi-generation technologies are intentionally configured into an integrated operating system versus operating independently and "integrated" via the power grid. For tightly coupled integrated systems, the multiple output streams represent different system optimization operating alternatives, which included the use of heat and/or electricity for hydrogen generation, water purification, pumping or desalination, and industrial processes. Innovation in this space continued to evolve; more complex thinking started to open up creative planning and analyses, and develop opportunities to introduce multi-input/multi-output hybrid configurations that offered resource resiliency with multiple product and revenue streams.[76] But, these systems also introduced increased complexity in system design engineering and operating optimization requirements at the physical level (as opposed to financial hedging). The momentum for these integrated systems is building today under the rubric of low-carbon industrial clusters. In the U.S., this is particularly motivated by the U.S. Department of Energy's call for hydrogen hubs.[77] In Europe, it falls under multiple areas of motivation including low-carbon hydrogen and low-carbon industrial transformation.[78]

The strong combination of factors during these two decades led to explosive growth in renewables, the knowledge to use them, the economic attractiveness of doing so, and the social and institutional willingness to support change. With this momentum, renewables are poised to enter a new phase, bringing along enormous opportunity and responsibility. Let us explore what the landscape looks like.

Snapshots from 2000–2020

The Energy Policy Act of 2005 creates a 30% federal investment tax credit (ITC)[79] for residential and commercial solar energy systems. The credit is extended in 2006, 2008, and 2015.

Danish wind generation grows from a capacity of a few 100 MW in the early 1990s to 2.5 GW in 2002 and 6 GW in 2020.

Chinese production of PV skyrocketed from virtually zero in 2000 to 2.5 GW in 2007 to 50 GW in 2016.

Japan's market share of worldwide PV production drops from 40% in 2000 to less than 5% in 2016. China's market share grows to more than 50% by 2016 and 80% by 2020.

The original Renewable Energy Directive, adopted on April 23 2009 (Directive 2009/28/EC,[80] repealing Directives 2001/77/EC[81] and 2003/30/EC),[82] established that a mandatory 20% share of EU energy consumption must come from renewable energy sources by 2020.

The U.S. residential solar market installation amounted to a capacity of over 2 GW in one year for the first time in 2015. The total U.S. installed market surpassed a capacity of 20 GW.[83]

In April 2016, the U.S. installed its one-millionth PV array.[84] The U.S. installed a capacity of 14,625 MW[85] over the year, a 95% increase over the record-breaking capacity of 7493 MW of 2015. Solar power is also ranked as the No. 1 source of new electricity-generating capacity on an annual basis for the first time ever. A new megawatt of solar PV came online every 36 minutes in 2016.

The installed cost of solar power fell to record lows in 2017. The total installed system cost declined in Q1[86] to $2.80/Wdc for residential, $1.85/Wdc

(Continued)

for commercial, $1.03/Wdc for fixed-tilt utility, and $1.11/Wdc for single-axis tracking utility, a 6–29% decrease from the previous year, depending on the market. The LCOE for fixed-tilt utility fell to between 5.0 and 6.6 cents/kWh and 4.4 and 6.1 cents/kWh for single-axis tracking utility, meeting the SunShot utility-scale goal three years early.

138 GW of PV was installed in 2020 globally.

The cumulative installed capacity for PV at the end of 2020 reached 760 GW.

The competitive tender (auction) prices for utility PV were at or below $20/MWhr.

New wind capacity of 93 GW was installed in 2020, dominated by China and the U.S.

The cumulative installed wind capacity was 743 GW.

Chapter 4
Phase IV: *Dominance* – From Petro to Electro (2020+)

Key Themes

- The convergence of factors

- New value propositions

- New services

- The end of energy poverty

- The new geopolitics

- Framework for the next decades: Five forces of business innovation in a decarbonized world, after COVID-19

Imagine the world in 2050…

It is 2050. The world is free of energy poverty and there are sufficient energy services to provide for the needs for all. There is clean water, electricity, fast digital communications, well-insulated and comfortable housing, piped water and sanitation, and abundantly productive agriculture which provides enough food to feed everyone a comfortable and healthy diet. And, we are doing it all, everywhere, without polluting the earth, the oceans, or the atmosphere. How has this come to be? We have learned to harness the renewable energy sources around the world. Technologies for capturing and utilizing or storing CO_2 from fossil fuels have advanced sufficiently to play a meaningful role

(Continued)

(Continued)

in the energy economy along with continued operation but different uses of an aging nuclear fleet. Within this complex web of energy supplies and end uses, renewables play a defining role. It is their era for dominance.

Enabled by a complex mix of digital controls, distributed energy systems, integrated energy systems, and interconnected power grids all transmitting clean electrons, those clean electrons are used to create and support clean chemicals and fuels which in turn go into nearly every product that we so enjoy. Steel, aluminum, and concrete are all made using processes that leverage this renewable electricity future. Plastics – and other consumable materials – come from a full reuse and recycle industry. There is no more plastic waste. Energy is not a limitation, but an enabler of economic advancement, education, health…

The world described in the textbox above involves a fundamentally new energy economy. The transformation of today's energy economy to one that is fully sustainable, low-carbon, just, reliable, and affordable will require enormous change. Phase IV below offers a set of insights that form the foundation for this worldwide transformation.

Getting there from here…

Reflecting on the first three Phases offers a perspective from which to look forward. We have now entered Phase IV. A combination of the six factors has created an enabling environment for profound change. Technological progress has and will continue to be exceptionally productive, as shown in Figure 1. One can foresee a fascinating future where the interplay between data, computing, artificial intelligence, communications, energy technologies, sensing, smart design, and innovative science comes together to provide zero-carbon solutions across economies around the world. The cumulative knowledge that has been gained over the last few decades is coming to fruition in ways that could not have been imagined. The advancements of artificial intelligence in combination with distributed computing and communications and technologies will continue to enable unprecedented solutions to be created, commercialized, and adopted at scale.

Dominance

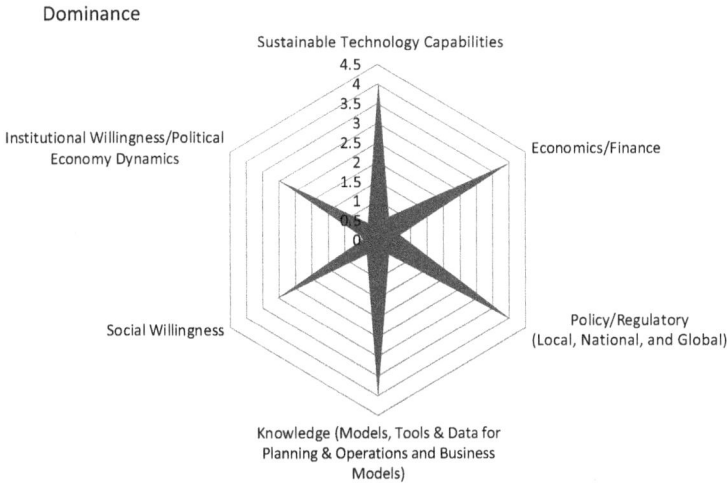

Figure 1.   A changing set of factors for Phase IV.

While neither ideal nor uniform around the world, there is an increase in national, local, and individual attention to making substantial changes in our energy system and addressing global climate change. Decarbonizing the power system and energy economies around the world while also providing for energy access is a fundamental tenet of these strategies. The economics are increasingly compelling as renewable power generation is the least cost option in nearly every country in the world compared to fossil fuel costs as well as nuclear costs, including capital and operating expenses. Not all countries, however, are as richly endowed in renewable resources as the U.S., Brazil, Uruguay, Denmark, and others. This implies complex regional development with all of the accompanying economic and geopolitical factors. Institutionally, there is growing support in many but not all sectors or institutions across economies. Further, many corporations have committed to 100% renewable, 100% clean, or net-zero transitions. Others have committed their full suite of product lines to be aligned with net-zero energy companies such as electric vehicles (with the simultaneous decarbonization of the power system). Many investors, financing institutions, leading private equity groups, commercial investment banking houses, central banks, and multilateral development banks have announced climate-aligned investment portfolios. Policies are being developed in multiple contexts, ranging from nationally determined contributions (NDCs) through to local jurisdictions that are committed to supporting clean-energy transitions.[87] However, these are not uniform and do not cover every jurisdiction or every country. In summary, the combination of factors is strongly in favor of a transition to a renewable-energy-dominated low-carbon future.

But, the pathway to get there will be rocky and turbulent. It will not be uniform and likely not fast enough.

Forecasted statistics for this era are even more stunning than in Phase III. Global analysis produced by the International Energy Agency paints a picture of 2050 where the world economy is nearly 40% higher (with 2 billion more people), but overall energy consumption is 7% lower than what it is today. This implies an incredible amount of energy efficiency and rethinking of system efficiencies going forward. In 2050, 50% of the total energy demand will be from electricity and renewables will supply nearly 90% of global electricity, with nearly 70% of that coming from wind and solar alone. On the way to 2050, solar installations are projected to grow to 600 GW per year by 2030. More recent indicators from industry are that the solar sector may indeed be at a production capacity of one terawatt by as early as 2025. That may provide a ray of hope, but not without caution. On the demand side, EV sales will rocket to over 50% of all vehicle sales by 2030 and will be the dominant mode by 2050. Electrification of mobility more broadly will be extended from scooters to motorcycles to light-duty vehicles all the way to heavy-duty trucking. Aviation will see increased utilization of small, electrified aircraft for short- and medium-haul routes in many countries. Renewable-energy-derived fuels plus those from anthropogenic sources combined with carbon capture and storage will continue to support long-haul aviation. Marine shipping will have mostly transitioned to usage of ammonia, methanol, or other fuels. Chemical supply chains and industrial processes will have converted to sustainable supply chains and electrification or clean hydrogen-based processes. Overall oil demand may be as low as 20 million barrels per day (MBD) down from ~100 MBD that was experienced over the past multiple decades. Such an energy transition requires a substantial shift from petroleum and fossil fuel ongoing supply chains to investments in critical minerals and materials to create the renewable-electricity-based energy ecosystem. Thus, mining will also have transformed to low-carbon machinery and minimal environmental footprints.

In this section, the interplay across the technology revolution and the factors affecting the speed and scale of our energy transition to a renewable-energy-dominated future are further explored.

The Convergence of Factors

Sustainable technology capabilities

IT and ET (Information technologies and energy technologies)

Information technology (IT) and energy technology (ET). Or is it IoET? Nearly everything that touches our lives today has been influenced by the transformation of IT, cellular telephony, and the Internet. Interestingly, most people, because these technologies are consistently interacting with their cognitive processes by providing information, answering questions, providing a basis for learning, or providing a basis for commerce and entertainment, are keenly aware of the influence and power of IT. What is perhaps more interesting from an energy perspective is the lack of awareness of energy despite it being required for everything to do with IT. Servers would not run without electricity, cell phones will not connect without power, and the cell network cannot work without power. But, few think about power or energy. They do of course when the lights go out or their phone needs charging. This perhaps is both a compliment and a challenge introduced by the engineers and operators who maintain a reliable power system in most locations (but not all). This is very different from location to location around the world and certainly for those without access to modern energy. However, the integration of IT and ET is pervasive in the provision of modern energy around the world. The integration of IT and ET extends broadly and deeply across sensors, distributed computing, AI and machine learning (ML), cloud-based computing, and resources all connected to energy infrastructures and devices.

Looking at the evolution of the provision of electricity-based services, lighting, communications, and power for small industries and rural communities around the world, the expansion and availability of cellular telephony has been game-changing. Not only do the towers and repeater stations provide anchor tenants for small systems that can then be leveraged to provide energy services to local residences, but the availability of data and communications has also transformed such systems that can be operated, monitored, and paid for via telephony. Integrating a cellular data stream to a controller of a small solar-battery-based hybrid system is one example. It used to be that time-based controllers were used where one had to insert a coin to operate a timer to provide power. Today, payment systems are facilitated through SMS or other data messages for online accounts, linked through financial accounts or other payment systems. Additionally, the controller can monitor the health state of the system, report problems back, and call and dispatch technicians if needed.

Connecting distributed and decentralized assets through digital communication also enables the aggregation of loads, as noted, that can provide value-added services to a bulk power system. The more recent lexicon for aggregation is captured within the concept of Virtual Power Plants. These allow the services of aggregated assets to participate in power markets. Here, for example, consider controllable hot water heaters. Hot water heaters usually include a storage tank which is thermally insulated typically with a time constant of many minutes to hours. Classic controllers sense the temperature in the tank and compute a differential between that and a set point. If that differential is greater than a few degrees, typically, the controller will turn on and call for power to reduce the differential. In classically designed systems, this operates 24 hrs a day, 365 days a year. Now, consider the alternative in which the controller is connected to a Wi-Fi network or a cellular network and can receive and follow signals from an aggregation control center. In a situation where the system operator needs to reduce the load (for example, when it does not have enough reserve power generation to meet the load, or when it may be more cost-effective to reduce the load than call on expensive or high-emitting generators), it sends a request or a signal to the distribution control center. That center then communicates to thousands or tens of thousands of hot water heaters. It asks them to remain warm, but not continue heating, for the next several minutes or potentially hours as requested to allow the system operator to support reliable, low-cost operations, and perhaps carbon emissions goals. The stored hot water in the tanks remains hot, the power system meets its goals, and the consumer may be paid for their "service" and everyone saves money. Does this really make a difference? On average, electric hot water heaters pull 4.5 kW each. That means for each thousand hot water heaters which can be turned off for a given period of time, the system operator reduces the generation requirement by 4.5 MW. 10,000 hot water heaters amount to 4.5 GW, or approximately $4.5 billion in capital investment.

Similarly, air conditioners and room cooling represent an enormous and growing opportunity for aggregated demand response or load shifting. Today, nearly 2 billion air conditioners are used around the world. More than 150 million are sold each year, predominantly in China and developing economies. According to the International Energy Agency report from 2018 on the Future of Cooling, projections indicate that nearly 6 billion units or more are anticipated to be sold by 2050, given that less than 10% of the population in developing economies and hot and humid climates currently have air conditioners.[88] A few elements are important regarding this projected explosive growth in air-conditioning: demand and peak load. First, air conditioners must be as efficient as possible. This impacts overall capacity needs and energy needs. Second, where possible, align air-conditioning use with availability of renewable resources. This is relatively straightforward in sunny, hot, humid climates during the day, but does not address evening and nighttime air-conditioning requirements. Third, consider digital controllers. Here, units that are digitally enabled could receive control signals, based on grid quality conditions, pricing, or greenhouse gas signals, to turn off the compressor and/or the fans in order to respond to the digital signal. For a sense of scale, consider a metropolitan area of 4 million households, a relatively modest-sized city in many countries. With an air-conditioning penetration or ownership rate of 100%, and an average load of 1 kW for a room air conditioner, potential load shifting or peak reduction of 4–10 GW is readily available. Fourth, consider alternative technologies. For example, alternative cooling solutions include district, or complex (campus) district, cooling in which the thermal mass of a nearby water body such as a lake or ocean is used to provide cooling. Another alternative utilizes price differentials and simple mechanical systems to provide cooling to the building. In this approach, ice is created at night using inexpensive and readily available power (or during the day if there is sufficient solar power generation). During the day when building temperature conditioning is needed, a highly efficient low-wattage fan blows air over the ice. Combined with a desiccant, dehumidified cold air can then be distributed throughout the building. Both these alternatives exist commercially today, but may not work for many locations in the world. Technically, they offer both energy use reductions and peak load reductions of about 50% or greater. Dense urban environments, or those not located near a body of water, represent significant challenges for cooling.

Another example on the cusp of significant change is electric mobility. As of 2022, nearly every light-duty vehicle car manufacturer has released or committed to releasing electric vehicles, and possibly changing their complete fleet to electric by 2030 or at latest by 2050. The rapid expansion of electric vehicles presents an enormous opportunity to reduce mobility greenhouse gas emissions as long as the grid power for charging is decarbonized. However, charging of electric vehicles may also be problematic, or potentially not. Here, digitization, smart controls, creative pricing and tariff schemes, and AI all come into play. Without intelligence, vehicles may only report the state of the battery. However, with intelligence, which learns driving patterns, vehicles can also report distances to a given state of charge, as well as recommend nearby charging stations, times for availability of charging, pricing, and greenhouse gas footprint, even making an automatic reservation. During the introduction of electric vehicles, there was a myth that the range a vehicle is able to drive was the principal criterion for purchasing. For some, this may have been a consideration. For others and in particular the billions who live in dense urban environments, it is less of a consideration. A second myth was also pervasive. It was believed that people would want their vehicles 100% charged every day. That implies that the EV owner would use the vehicle, perhaps to commute to and from work, and come home from work and immediately plug the vehicle in to recharge it. This would add to an already strong evening peak demand on top of the lighting, cooking, and air-conditioning or heating loads. Schemes were created combining digital communications and controls to delay the charging to the nighttime. This helped keep the evening peak load manageable within the existing available generation. Additionally, it was found that many EV owners do not plug their vehicle in each night because they know, either from a previous driving session or via a smart app on their phone, what the state of charge is on their car and decide if the car needs to charge or not. This is very similar to the behavior of nearly every owner of an internal combustion engine light-duty vehicle who refills the gasoline tank only when it is nearly empty. Of course, time frames can be different depending upon the level of the electric vehicle charger. Normally, refilling a gasoline tank takes approximately 5 minutes. Sometimes, vehicle charging can be as short as 5 or 10 minutes or as long as multiple hours depending on the strength of the charger. Interestingly, new super-capacitor-based batteries are being introduced into the marketplace which could reduce charging times to at short as 30 seconds.

Let's dive even further into the integration of IT and ET. Consider, for example, the futuristic scenarios envisioned under the rubric of "autonomous energy systems." Here, a number of forward-thinking engineers realize the challenge and opportunity of a highly distributed, decentralized, digitized, diverse (technologically), and decarbonized power system.[89] To get a picture of this, imagine an electrified house in which all lighting, heating, cooling, cooking, and other appliances are electric and addressable at the individual level, that is, each individual light bulb, for example, or room of lights, every plug, and every appliance. This is not so far off from current capabilities given the Wi-Fi-enabled auditory controls available today. However, in this house, they also have solar PV panels, each with an individual micro inverter, combined with a battery storage system for the house, and two or more electric vehicle chargers. This house of course is one of many in a neighborhood in which they share a community-based micro grid that could be comprised of wind turbines, solar, hydro, and additional storage technologies such as hydrogen, made by an integrated electrolyzer, compressor, storage system, etc. The micro grid has its own control algorithm in which it learns and senses the weather, the household energy use patterns, traffic patterns, and thus EV charging requirements and the state of the bulk power grid to which it is connected. Questions arise around this configuration: Should each household configuration self-optimize? Or should some of the houses cooperate to optimize and be part of a cooperative set of configurations for the community? Should those houses contribute services to the micro grid or should the micro grid just be supportive of household demand and play a bridging role between the community and the bulk grid? At what level should control of individual devices be conducted – at the household, at the community, or at the sub-cooperative level? Might there be an override that gives the bulk power system priority controls over some of the assets (micro grid, houses, or individual devices) to potentially step into the operational chain? Where are the cyber threats if everything is connected to the Internet? How can this small-energy ecosystem be protected from cyber hacking? What are the relative costs and benefits of these different configurations?

It was this series of questions that prompted the line of inquiry that began the development of a fundamental series of algorithms for distributed microcellular control of various assets. Those assets could be wind turbines within a wind farm or perhaps individual devices within a house, or multiple households within a community. None of this would be possible without the direct integration of IT and ET, as data sensing, computing, and controls are central to enabling these possibilities.

Two examples of this distributed independent but coordinated control approach initially showed potential high value. The first example is microcellular control in a wind farm. For the past multiple decades, wind farms were, in hindsight, not very intelligently designed. Engineers placed the same type and height of turbine in a linear row to take advantage of the predominant wind direction and wind speed. Those turbines could pitch and yaw to follow the wind. But, turbulence has been studied since the mid-1800s. While the first turbine in a row experiences nominally uniform wind conditions, its interaction with the wind creates turbulence for the next turbine in that row and so on. Depicted in a classic photograph of turbulence in a fog bank moving through an offshore wind farm at Horns Rev 1 in Denmark in 2008, engineers realized that turbines in rows $n \geq 2$ experienced the "wake effect" from the prior turbines. The turbulence not only reduced the power output but also induced stress and increased degradation of the blades which were vibrating at seemingly chaotic frequencies and intensities.[90] For the next 10 years, engineers and scientists focused on new approaches, including 3D wind farm designs and the use of microcellular control in which cells of 3–5 turbines could optimize, given their spatially specific conditions while also taking into account the potential impacts that positioning would have on nearby microcells. By creating this integrated network of microcells, engineers were able to show improved outputs of a given wind farm of about 5%. Combining these advanced control approaches with 3D design optimization offers even greater potential for improving output, reliability, and cost-effectiveness.[91]

Three-dimensional wind plant designs are now enabled by advances in computing, AI and ML, and computational power in order to solve the complex flow equations. Today, there is a full suite of available modeling tools created by leading institutes from around the world. These tools support real-time control capabilities through sophisticated blade designs, balance-of-systems designs, and wind plant designs and operations. Looking toward the interface between atmospheric physics, wind power, and computing, there is the recently developed ExaWind, which is an open-source platform that ranges from the micron-scale physics to the kilometer-scale physics of a wind farm.[92] While targeted for use on super computers (e.g., Petaflop to Exaflop), these codes enable deeper learning, and through extensive validation, present the ability to translate those codes into faster representations of the complex physics that are sufficient for non-exascale computer applications. This allows effective representation of these phenomena in computational models that can run on smaller computers, which can be incorporated into decision-making analyses and methods more readily.

The second example of distributed, digitized controls includes large-scale co-optimization of distributed energy resources, electricity production, and distributed loads all as a function of time. Here, engineers were able to create a digital twin of the San Francisco Bay Area which included 10 million devices. Following complementary optimization methods, the placement of solar panels – some in combination with batteries depending upon the building load profiles and local grid conditions – was identified for 4.3 million customers. Additionally, traffic patterns were simulated based on AI-enabled algorithms using terabytes of existing data and combined with electric vehicle driving patterns, battery use, and charging requirements, all informed by AI-enabled algorithms of existing datasets. Options considered individual driving and mobility as service use patterns. The simulation was then able to co-simulate an operating system at one second resolution that included dynamic cloud cover, correlated solar output, and (where needed) optimized battery output, in order to meet the simulated building loads as well as electric vehicle charging loads for ~400,000 buildings/solar/battery systems and 10,000 EVs.[93] Here, the interplay with AI opens up a potentially broad landscape for control optimization. While spatially the initial control approach was toward microcells in communication with the next-nearest neighbors and perhaps even a higher-level control algorithm, the approach to those controls and effective communication could be subject to a number of approaches that are being adopted from AI. This includes concepts of consensus-based controls, beehive optimization, and ant colony optimization. This opens up an incredible landscape for creativity, continued learning, and value creation for both businesses and individuals, as well as society.

IT is also central to utility operations, from billing and payment systems to the complex controls of a power-generating station and a command and control center for system operations. Systems which used to be mechanical have nearly all converted to digital. On the generation side, solar, wind, and batteries all contribute to the grid via a power inverter. This fundamental building block that converts DC power to AC is fully digital. Over the past 10 years, capabilities have expanded dramatically. This includes Wi-Fi or cellular-based auto upgrade functions to very sophisticated sensing and controls of voltage and frequency; even today, inertia is sensed in the power system. The evolution of digitization and high-performing power electronics is enabling a fundamental transformation in the perception and capabilities of renewable-based generators. What were once considered problematic for the grid are now considered valuable grid assets. Here, for example, large-scale systems can provide services such as up-ramping, down-ramping, voltage support, frequency support, and auto generation control in which the amount of power provided to the grid follows the requested amount from the system operator very closely.

One of the first experiments of this capability was reported for a solar plant in California. It was enabled to provide auto generation, showing that the solar plant could follow a signal within four seconds and provide power with ~90% accuracy of the requested amount vs 63% for a natural gas combustion turbine, which is generally considered the fastest response technology. Today, voltage support, frequency support, auto generation control, and multiple other features are all embedded in the best-in-class inverters.[94] These capabilities provide critical resources for situations when there is an imbalance in the grid, e.g., when the frequency or voltage is out of tolerance. Further, cutting-edge research is focused on creating capabilities for "grid formation" from inverters; that is, adding power while maintaining voltage and frequency to meet the full load for a system configuration.

Forecasting

Forecasting of load, resources (for power), and consumer choices/behavior by accounting for the day, weather, activities (such as major sporting events), etc., began to emerge as a more complete set of practices in the mid-2000s. What exactly does this mean and where is it going? Perhaps a step back is in order. Traditionally, power system forecasting relies upon historic load curves that are used in order to anticipate future or real-time anticipated loads. Such load curves might be based on utilization of the prior day, the prior week, or a day under similar weather or other conditions, such as holidays or, for example, in Brazil, days in which the national team is playing in the World Cup (and thus many people take off work and watch the match on television, affecting the load profiles around the country). These load curves gave system operators the necessary information to call upon generators in a different part of their system to meet the load across the full system including imports and exports from interconnected grids. The fundamental principle of understanding anticipated load and supply at any given time is essential in order to ensure real-time matching. Today, and going forward, the tools and the options available are much more powerful than just a few days of prior load profiles and the ability to turn on or off generators, import or export power.

Forecasting today can incorporate the understanding of near-term information to be able to make real-time (or near-real-time) operating decisions. This contrasts with longer-term planning which will be discussed later. There are three aspects of forecasting to consider. The first aspect is energy operational decisions made within ideally a few days of forward information. From any energy system perspective – whether addressing the production of oil and natural gas or delivery, processing the storage and use of chemicals, or the anticipated use of power for residential, commercial, or industrial uses, including vehicle charging or even electric air mobility – all aspects must be addressed to accurately forecast the anticipated demand and supply of energy. Need here refers to load for a power system, perhaps barrels of oil, or million Btus of natural gas. The second aspect is understanding supply, that is, the availability of key inputs whether or not that is for product processing, heat, or the production of power. For the production of power with increased penetration of renewables, the weather-dependent variability becomes increasingly important. Advances in using sophisticated weather models in combination with AI-enabled downscaling techniques and plant production software are enabling power utilities to have significantly improved confidence in the day-ahead to hour-ahead availability of power from renewable generation facilities.

The third aspect is the interplay between load, supply, operations, knowledge, and markets. In nearly every energy market ranging from the oil and gas markets to petrochemical to power, each step in the conversion of a primary energy source into useful products and services involves optimization of supply, production, and economics. In the power markets, multiple forecasting components are working simultaneously. In markets that have both day-ahead bidding and near-real-time market closure, forecasting and economic analysis determine the participating assets in the day-ahead market. However, as it advances closer to the real time, some of those assets may have changed under their economic position becoming operationally unavailable or the weather patterns may have changed, affecting the output of either solar or wind plants. Because all assets in the power system have different response times, the system operator needs to provide signaling at the appropriate time in order to ensure that the system has a sufficient balance of load-and-supply assets to run reliably. And, renewable assets can respond to different market signals for the provision of different services in those markets. This description holds for principally direct grid-connected power plants. However, it also applies to grid-connected hybrid systems and potentially to aggregated distributed energy resources. Having advanced forecasting which is enabled through learning certain data algorithms is a fundamental tenet for today's energy systems as well as those of our future.

Sensors/Data processing

A revolution in sensors and the data they provide has also been underway for nearly a decade. Today, there is more information, and more data, than can be used effectively without continuing to innovate. To illustrate with one example, consider PG&E, the California utility which has unfortunately experienced disastrous fires over the past few years. PG&E operates 1,300 weather stations providing real-time information of wind, wind gust, temperature, and humidity, including real-time video streaming. All the information is streamed to a real-time weather site and also used in combination with satellite information that has fire information that is as up-to-date as possible. However, so much more could be done. Consider potentially linking the real-time weather stations with day-ahead forecast to improve the predictability and insights for the utility and customers.

Cybersecurity

Our evolving, growing, increasingly digitized, connected, and distributed energy systems are at risk of increasingly complex cybersecurity incidents. The more devices that are connected to the Internet and connected through control algorithms, the more the breadth and depth of "threat vectors" increase, i.e., the ability to hack into the system through a vulnerable device and gain access to affect targeted or system-wide operations of a highly electrified energy system. Here, consider the electrified interconnected PV, battery, building, EV charger, or grid city scale that was described earlier in which each device in the house is connected to the Internet, along with the presence of a home energy management system, microcellular controls, community-wide systems and controls, and bulk system controls. In this multi-layered system of systems, each individual component in and of itself should be secure, and the interface among the systems must also be secure.

There are four primary strategies being pursued to cover the landscape of exponential increases in interconnected energy devices. The first strategy involves a module, which is a set of capabilities designed to protect legacy protocols and aging devices against cybersecurity issues with minimal obstruction to normal operations. It is an add-on set of protocols that have been developed and certified using public key encryption.

The second approach involves incorporating Blockchain for security and energy management. While still in the early stages of development for energy cybersecurity uses, Blockchain is well established in multiple other sectors, providing assured identification of origin and supply chains for such items as diamonds, some metals, and financial assets. Here, the concept is that Blockchain identifiers could be affiliated with energy assets and through that secure identifier could control and deliver commands.

The third strategy revolves around leveraging 5G communication and its capabilities with interconnected energy assets. Of course, the requirement is that 5G be available. One of the key features within 5G is the ability of software to find a network. This allows smart engineers and software packages to dynamically define networks or "slice" them dynamically among devices that need secure communications in real time. Here, perhaps it is important to remember that interconnected devices and electric systems have traditionally had dedicated networks to provide near-real-time control capabilities in order to enable real-time balancing of supply and demand. 5G communication offers the ability to rapidly reconfigure networks among devices in combination with massive machine-to-machine learning and ultra-reliable low-latency communication. The low latency is important for the real-time balance of the power system. Use cases and standards are being developed through the National Electric Sector Cybersecurity Organizational Resource Working Group. Advanced networks are not without concern. Some have expressed concern over the energy intensity of the network frequency; however, the science is relatively clear that this is not harmful. Nevertheless, long-term studies have yet to be conducted. For personal security, identity theft remains a ubiquitous concern for all of the IoET applications. Lastly, looking forward, consider the fact that the next generation of technologies, e.g., 6G and beyond, is in development which offers even further enhancement of capabilities, enabling a connected, clean-energy future.

The fourth approach involves leveraging other relatively ubiquitous infrastructure, at least in Europe and the U.S., in many cities around the world. This other infrastructure is the cable TV infrastructure. Cable TV networks carry information that can be used in combination with GPS identifier information, weather information, electric grid models and information from smart chargers, home energy management systems, and distributed energy management systems to inform system operators of situational anomalies. Such anomalies may be local power outages or local threat attacks that can be identified through the comparative analysis of power sector and cable TV sector information.

Cybersecurity has a critical role to play today and even more so going forward. These few examples touch only lightly upon the various approaches that are in development. Continued focus on the evolution of software and hardware programs is anticipated, along with a combination of protocols and leveraging communications infrastructure, to continue to advance cyber-resilient solutions. The solutions will play out at different scales around the world. The process described is principally oriented toward highly electrified, distributed, digital, decarbonized environments. In many locations where only solar home systems or mini grids provide basic energy services, cyber threats are less prevalent. However, in every location where there is a large interconnected grid with multiple assets connected, cybersecurity remains a core security and resiliency function of the power generators and system operators.

AI/ML for energy

As mentioned earlier, AI is being infused in various parts of the energy economy. As previously mentioned, electric vehicle use, state of charge, intelligence to support timely charging, etc., are all enabled by AI algorithms and on-board computing. Households, commercial buildings, and industrial processes can be increasingly automated with self-learning algorithms. This is already relatively ubiquitously applied to equipment failures and preventative maintenance in large industrial process facilities. In residential applications, the smart thermostat stands out as the epitome of this application; however, integrated home energy management systems are evolving to do much more, including managing thermal comfort, hot water, and other appliances throughout the house as well as EV charging. From a larger systems perspective, AI is being used to combine occupant data, use data, weather patterns, renewable generation patterns, carbon emissions, and system pricing, all to offer the ability to self-optimize at a building level, micro cell level, or micro grid level. Hundreds of AI algorithms have been tested and applied to hybrid system optimization routines. Incorporating multiple other factors than just power system characteristics seems to offer enhanced capabilities to meet customer needs through reliable services at reduced cost or reduced emissions.

Some less obvious uses of AI approaches include developing methods for downscaling of global climate modeling results to useful geospatial and temporal scales for energy planning. Here, climate scenario datasets are available for training at the terabyte scale. Using those trained algorithms, researchers can then apply them to future climate scenarios through 2100 or beyond. For example, researchers have shown that climate data of relatively coarse scale, 200 km × 200 km, can be downscaled to 2-km resolution which is sufficient for energy system planning and in particular for calculating spatial and temporal energy production from wind and solar in future years under different climate scenarios. This is an incredibly important breakthrough as nearly all energy planning for the future uses historical datasets as representative years with probability distributions that reflect the variability of historic patterns. However, it is now firmly established that climate and extreme weather, as well as non-extreme weather, are changing. This is epitomized by the increasing frequency and intensity of hurricanes, monsoons, or long-term fluctuations in global ocean circulation, all of which have affected the median and the distributions of wind patterns on the Indian subcontinent. Those long-term fluctuations which may occur over 20 or more years may be significant to obtain the output of wind generation plants or solar generation plants which are intended to be in service for 20–50 years. Other examples include volcanic eruptions which might reduce solar radiation and thus the output of solar PV significantly for considerable amounts of time, not to mention their impact on the greenhouse gas emissions inventory as well as solar flux hitting the earth. Thus, utilizing future data to be able to effectively plan for resilient, zero-carbon energy systems is becoming increasingly important.

Humans in the loop

There has been a pervasive myth in the energy innovation sector that more information will lead to improved personal and household energy decisions, particularly in developed countries. However, personal interest in energy varies with availability and cost of energy. In the years and decades where energy was abundant and cheap, most people did not pay attention to it. In fact, many people do not know where electricity comes from besides the outlet on the wall or the light switch. With average bills of 3–5 U.S. dollars per day, less than a cup of coffee in the U.S., one can see why energy was not of interest. Gasoline and gasoline prices draw more attention given the advertisements among news and radio stations, news coverage, and awareness of spending in somewhat larger increments. Early experiments to provide more information to homeowners, such as through the Google energy meter, proved to not be very sticky. People did not care and do not want to take the time to actively manage their appliances and energy use. That led to the introduction of smart devices and home energy management systems. In the future, further extensions of personal learning devices that help manage our energy and climate footprint, all while being seamlessly invisible, will continue to emerge.

Energy enables four fundamental aspects of our lives: comfort, connectivity, mobility, and goods. Consider our personal temperature management. Nearly all of us control this today by setting a room temperature thermostat. Of course, we individually manage the clothes we wear. But rarely, except for a few who use high-end automobiles, do we have the opportunity to individually manage our interface with chairs, beds, or couches. In the future, the features of high-end

automobile seats, which include perforated seating for airflow cooling and embedded heating elements for warmth, could be increasingly available in chairs throughout your house, office, meeting rooms, and hotels, and even couches and beds. Those devices could learn and pre-condition for your personal preferences, know your whereabouts, or respond to your body temperature given a signal from your smart watch. The room itself would be at a different temperature to optimize for energy efficiency, pricing, or GHG emissions.

Whether a phone call, the Internet, or a video chat, all types of connectivity rely on electricity to power the devices and the networks. The energy implications of the connected world are relatively significant. Some estimate that data centers, cell phone networks, and storage of explosively growing data will consume 10–25% of global energy demand in the not-too-distant future. This reliance on energy, and its environmental footprint, is well recognized in the broader IT industry by companies such as Google, Apple, Microsoft, and Amazon. The global mobile telephony industry is also committed to a sustainable pathway. It is a good thing that the industry is committed to decarbonizing its footprint and its services. We, the users, may be committed to reducing our own environmental footprint and the actions that we can make. Phones and computers play a very different role in our lives, and we have little if any direct influence and certainly no control over the energy and environmental footprint of the services that they provide. In 2016, an interesting experiment was conducted in this realm in which customers were asked if they would prefer to download a movie with a lower environmental footprint. Around 42% of the respondents said they would do so if offered the option. This was one of four experiments conducted at that time to evaluate whether or not firms could lower operating costs and environmental footprint, and provide differentiated products and services to customers. One of the other experiments explored if customers would choose to offset the carbon footprint of a shared mobility ride. More than 80% of the customers stated that they would do so, if that box was checked as the default on the page and the price was as little as 2 cents or up to $0.20. An additional option was then provided that said, "Please do this for all future rides." A significant portion of respondents also indicated that as an ongoing option.[a] A few of the road service

---

a These four experiments and their findings are described in more detail in Isley *et al.*[95]

companies actually began to implement this before the explosion of EVs in their fleet. It was then discontinued. However, algorithms have been developed for more ubiquitous applications such as Google Maps. Google Maps now offers a Greenleaf indicator for the route with the lowest environmental footprint and makes that recommendation.

Meeting in person, e.g., "in-person connectivity," is deeply intertwined with the availability of energy, vehicles, mass transit, and other mode options. Even very short distances which are served by human energy have an energy and environmental footprint. We expend our energy walking, running, or biking. That energy is derived from food and water, both of which have related energy requirements and environmental footprints. While we will address the energy–food–water–environment nexus separately, let us first look at mechanized mobility and its future in the net-zero energy economy.

Mobility choice is a complex decision dynamic for nearly everyone that includes considerations of distance, time, convenience, safety, economics, mode choice, and environmental footprint.

For some, there are limited or no choice options. That is, there may not be mass transit nearby, nor safe lanes to walk or scooters to ride. For others, choices may be limited by economic circumstances. For the hundreds of millions, if not billions in the coming decades, who will be predominantly living in urban environments, there will be an increasing suite of options to choose from. We see examples of this already, including the availability of scooters or bicycles to use on demand, shared ride options such as Uber or Lyft, and of course traditional buses, trains, and subways. Concepts of personalized air mobility via electric vertical take-off and landing vehicles (EVTOLs) in urban environments have been proposed for the future and are under development. For many who have been living in dense urban environments, a few of these mode choices have been available for decades. But, innovation and technology, payment mechanisms, and business models continue to expand the suite of options and offer pathways to lower environmental footprints as nearly all of the innovative options are electrified.

Beyond the urban corridor as well, innovation is coming. More than a dozen electric airplane manufacturers are in the pipeline for certification. Targeting regional short- and medium-haul routes, these fully electrified aircraft offer potentially game-changing environmental footprint reduction for aviation travel. Electrification of segments of the aviation industry implies significant change for training, flight operations, airports, and electricity requirement. Major training centers are already planning to convert 50% or more of their training fleet to electric aircraft. This will bring on a whole new generation of pilots who will understand the details of how to fly an electric aircraft. On the airport side, there is a significant movement among airports to also achieve net-zero greenhouse gas in Scope 1 and Scope 2 emissions, that is, their own operations and energy consumption. Airports are rapidly advancing to improve energy efficiency, electrify or convert to hydrogen ground operations, and support their staff and vendors to achieve zero-carbon emissions. Fueling of aircraft has normally been conducted by a third-party provider working on airport property. The expansion to hydrogen fueling implies a completely secondary fueling infrastructure, safety protocols, etc. Expansion to electric charging of aircraft is more complex. Typically, an airport owns its own electric distribution system. Few airports have on-site generation and most receive power from the local utility. Power loads at airports range from as low as a few hundred kilowatts for small regional airports to tens of megawatts, perhaps 100 MW, for a large multi-terminal airport in a major city. If electric aircraft require real-time refueling, that is, charging of batteries which remain in the aircraft, the necessary power demands may be as high as 4 MW per airplane. This implies that an airport may be a gigawatt-scale demand center with intense demand peaks during the day and low demand at night. Solving that delivery challenge, given the intense demands to retain as high an aircraft utilization capacity as possible, implies significant infrastructure investment. The technical solutions range from on-site generation and establishing a storage and distribution network to support aircraft charging to reinforcing the distribution grid to support GW-scale peak demands. At smaller regional airports, demands would grow potentially from a kilowatt scale to tens of megawatts. There are similar implications for infrastructure build-out, supply, storage, and delivery. Interestingly, some companies that are introducing small package delivery EVTOLs are proposing to build their own charging infrastructure based upon containerized solar PV and battery charging. They intend to implement a battery-swapping protocol. However, at this stage, for commercial passenger aircraft, such battery swapping would require certified personnel to perform the activity on each aircraft, at least in the U.S. It may be possible to train and certify a sufficient number of technicians coincident with the expansion of commercial electrified passenger air travel. Alternatively, one could pursue a regulatory change to allow non-certified personnel to perform a battery bank swap. There is no clear single winning strategy among these options. Hybrid configurations will likely also emerge. This is the basis for the next generation of GE engines, for example, hybridized turbines with an electric generator. Alternative approaches include hydrogen turbines combined with either electricity or sustainable aviation fuels (SAFs). Each of them has various challenges, cost implications, infrastructure build-out implications, and operating implications. It will only be through concerted multiparty dialogue, informed by deep technical and economic analysis, that we will find workable solutions.

In the previous sections, we have painted a picture of diversification of air mobility, most likely to play out in the U.S. and potentially a few other markets. However, in Europe, as well as for all transatlantic and other long-haul travel, sustainable low-carbon aviation will require what are called SAFs. Recognition for the need for low-carbon aviation fuel was prompted decades ago, and today, we have made significant progress toward delivering many billions of gallons of certified SAF per year. However, this is a very small percentage of the overall aviation fuel demand. Moreover, the delivery of nearly all of that SAF is either in America or in Europe. Multiple technical pathways are under development, leveraging biomass resources, used oils, used tires, or new approaches that derive SAFs from chemical processes that utilize clean hydrogen and carbon resources from either biomass or captured anthropogenic carbon dioxide. Attractive as a potential "drop-in" fuel (implying that the industry can use its existing infrastructure), currently certified SAF products have only been approved up to a certain blending limit because they have yet to achieve all of the technical requirements for 100% utilization.[b] In 2020 the U.S. government released its "SAF Grand Challenge." A coordinated effort of the Department of Energy, Department of Agriculture, and Department of Transportation, it lays out a pathway toward development and delivery of 3 billion gallons of SAF by 2030 and 35 billion gallons by 2050. Covering the full gamut ranging from R&D to standards development to process certification to market acceptance and scaling, the grand challenge provides a critical roadmap for SAFs in the U.S. Similar efforts are underway in Europe. And, while there is

---

b Fuels approved via ASTM D4054 and annexed under D7566 are considered synthetic blend components (SBCs). Current annexes require that SBCs be blended with conventional jet fuel. Once blended, the fuel meets ASTM D1655 criteria for aviation turbine fuel.

significant funding of approximately $15 billion to support energy innovation and the realization of the SAF Grand Challenge in the U.S., and comparable funding in Europe, providing global supplies of SAF represents an even broader challenge given the limited resource bases for local or regional production in many places around the world.

Hydrogen is also under consideration for aviation. It has been particularly emphasized in Europe, where there is less demand for short- and medium-haul flights, and Airbus has announced a major commitment to hydrogen-fueled aircraft. Here, too, we see a few new start-up companies entering the aircraft market coming up with concepts of hydrogen or hydrogen electric hybrids. Interestingly, fueling of aircraft with hydrogen falls under the current regulatory frameworks, in contrast to battery bank swapping as noted earlier.

Mobility vs connectivity: COVID-19 had a significant impact on mobility demand. During lockdowns, demand for vehicle mobility and air mobility dropped by 80%. However, the drop in mobility demand was short-lived. With vaccinations, and the lifting of quarantines, both vehicle demand and air mobility demand have rebounded to near-pre-pandemic levels. In the global work environment, the movement toward remote work for many, but not all, has implications for a slight reduction in mobility demand patterns. Daily commuting for many hundreds of millions of knowledge workers, programmers, and others who do computer or telephony-based work will likely remain an option for many years to come. However, daily commuting in the U.S. (based on a 2016 national transportation survey) only represented 15% of annual mobility demand, whereas 45% of daily trips are taken for shopping and errands and 27% of daily trips are social and recreational, such as visiting a friend. While a relatively small percentage of overall household mobility trips and associated emissions, the reductions in commuting amount to between 30–60 million hours per day, with an estimated economic impact of $90–$100 billion per year. And, people are more productive, healthier, and less stressed. It is anticipated that daily mobility patterns where remote work is possible will shift and remain over the longer term, but this will have limited impact on overall driving or mobility demand, particularly in emerging economies, as well as for many jobs that require physical presence. Social and recreational travel are very likely to grow substantially as income levels rise, implying greater importance of decarbonizing mobility.

Goods, the items we wear, use, consume, discard, and recycle, all have a considerable energy footprint – be that in the materials, processes, transportation, or preparation. In 2015, a team at NREL ran an interesting experiment to try to understand how more information could impact consumer choices relative to a set of criteria that reflected the person's individual preferences. Here, we gave people the option of price, environmental impact, convenience, and an applicable list of ingredients. We used augmented reality, in which software was developed such that a cell phone could recognize and differentiate among different bottled waters and then deliver information on top of a live camera image. Providing this additional information had a significant impact on the choice of bottled water and reducing the carbon footprint of the preferred and purchased items. In this particular case, it was an example where more information could actually help inform consumer choices that align with environmental values, and local, national, and global climate ambitions. In other instances where we used a similar technique to provide information for boxed cereal, it was interesting to note that even though that information on ingredients is available on the side of the box, providing it through the augmented reality app led consumers to choose healthier cereals than they would have otherwise. There may be health applications and potentially other uses that we have yet to conceive of or implement. Interestingly, today there is growing awareness of the carbon footprint of products, particularly related to importation – e.g., the carbon footprint of shipping.

Consciousness and awareness are very different in Europe where gasoline prices have averaged four times the U.S. price on a dollar-per-gallon or dollar-per-liter basis. Electricity is also more expensive in Europe. Energy consciousness rises in two particular instances. The first instance is when it is not available. In particular for power, system reliability and availability have accustomed people to electricity always being available. The loss of power from a major hurricane or a tornado is a significant event. Even a power outage from a downed tree or a nearby fire that may take out the electricity for hours to days brings renewed consciousness. Many who can afford it react by buying backup generators or installing photovoltaics and batteries.

The second situation is brought on by scarcity. Scarcity can be a consequence of war, embargoes, lack of supply chain availability, or other factors, in particular for countries, cities, or regions that are fully electrified. In 2022, spurred by a critical

energy crisis caused by the Russian invasion of Ukraine, Europeans responded with a wide variety of energy efficiency and energy supply measures. While many of these align with climate targets, some of them do not, such as restarting coal plants or building long-lasting infrastructure for LNG imports. However, energy security – and the provision of critical energy services through the winter to avoid loss of life and maintain optimum productivity – is the top priority. Scarcity is also evident in South Africa, where rolling brownouts have occurred on a regular basis for more than a year due to lack of sufficient supply. There is a long history to the current situation in South Africa that is an interesting lesson for us to reflect on. In the 1990s, when I started working there on behalf of the U.S. Department of Energy, Mandela had just been released from prison, and there was a focused program on delivering energy, water, and sanitation services for all. This was informed by a small cohort of energy analysts who evaluated the costs and benefits between off-grid systems, small-energy systems, water systems, and grid extension. The country was then long in generation capacity, i.e., they had gigawatts of excess. Over the following three decades, South Africa was unable to build sufficient new resources including those from within the very successful Renewable Energy-Independent Power Production (REIPP) program or even maintain some of the existing coal plants so that they were reliably operational. As demand grew and capacity did not, the country found itself in tight supply situations leading eventually to a short supply situation. Sadly, the financial situation, management, and governance have not expanded the REIPP program which took a very innovative approach, including the requirement for local jobs, local ownership, some local content, and guaranteed offtake contracts which would allow the renewable developers to have confidence in their financials and subsequently attract necessary investment.[96] Interestingly, in the mid-1990s, there was a misunderstanding among South African energy leaders that every gigawatt of renewables required a gigawatt of dispatchable "backup." This was addressed using examples from around the world in which that was not the case. But, it was not until the 2000s when sophisticated modeling of the South African grid enabled leaders to have confidence that they could successfully integrate tranches of renewable energy without building one-to-one dispatchable backup. Of course, we know today that one can operate up to 70% variable energy without significant changes in the capacity generation profiles, especially in temporally and spatially diverse power systems with significant amounts of renewables. Sadly, however, those myths persist in and among consultancies and many of the older trained engineers who are not aware of the state of the art.

The mobility landscape is also undergoing a fundamental transformation. While many of us are familiar with the Tesla story and the commitment from many manufacturers to produce a light-duty electric vehicle fleet and fully convert to that in the coming decades, electrification transformation is also occurring in medium- and heavy-duty vehicles with potential implications for transportation of goods, long-haul trucking, as well as aircraft as discussed previously. Electrification technologies are even infiltrating heavy machinery including mining trucks and construction vehicles. On the broader front of decarbonization of mobility, beyond SAFs, pathways for utilization of clean hydrogen and fuel cell vehicles or hybrid electric fuel cell vehicles continue to be developed. In the marine sector, there is an overall commitment to reduce the carbon footprint and potentially achieve net-zero emissions for maritime shipping. Strategies here focus on low-carbon fuels such as ammonia or methanol, and potentially hydrogen. All the strategies and transformations rely upon the decarbonized grid which will be predominantly based on renewable energy. Those clean electrons will be directly used either for battery charging and super capacitor charging, or for the creation of zero-carbon chemicals and fuels. For direct charging vehicles, as previously mentioned, there are significant implications for the power grid including supply as well as transmission and distribution. Take, for example, a supercharging station with multiple chargers. Each of those chargers may pull 250–350 kW, indicating a potential maximum peak load of 2.5–3.5 MW for 10 co-located chargers. Now, consider an airport with 10 rental car companies all wanting at least 20 fast charges. That equates to 200×350 kW or 70 MW of new load at a very localized location that was planned for and served via the build-out of local distribution grids for perhaps a MW of load. Given the need for wide distribution and availability of future charging stations, this has significant implications for the design and operations of distribution systems.

As battery performance improves, or as super-capacitor-based storage systems are adopted, some vehicle manufacturers are focusing on increasing the energy storage capacity of the vehicle, in other terms, extending the range the vehicle can be driven between charges. With energy capacities approaching 100–250 kWh for light-duty trucks and cars, and recognizing that new vehicle purchases are a major investment for most households, some manufacturers are proposing that vehicles be seen as a multipurpose asset for households. In this case, the vehicle would not only supply

mobility but also be used as a battery backup storage unit for household power. This is one application of the concept of vehicle to grid, or V2G. Under this concept, vehicles would not just be a load on the grid but could also provide power back into the grid and, if configured with appropriate additional technologies, may be able to provide other power services like aggregation.[97] That means, for example, that a company might offer a small reward to vehicle owners to use the energy stored in the vehicle to provide services to the distribution system operator by combining thousands of vehicles across the service territory simultaneously. This is a similar concept to that described previously for hot water heaters or air conditioners in terms of load management. Electric vehicles, on the other hand, offer the ability to not only be manageable loads but also potentially provide energy services to the grid.

Resiliency

Here, we want to differentiate between resiliency and reliability. Resiliency is the ability to withstand perturbations, either physical or non-physical. Reliability is the set of metrics and principles regarding the probabilistic uptime and availability of power. Under resiliency, which is increasingly a focus for system designers but also for component manufacturers and system operators, one has to address the complexity of both physical threats and cyber threats. We have outlined cybersecurity earlier. From the physical side, one can now much more clearly lay out a series of threats, either natural or human induced. On the natural side, with increasing intensity of extreme events, systems have to be rethought for resilient design for hurricanes, flooding, extreme freezing conditions, wildfires, drought, and potentially earthquakes in some regions. Techniques, including advanced simulation methodologies, allow for much more robust resiliency evaluation. The strategic issues at hand revolve around capital versus operating cost and lost economic productivity (or loss of lives, regretfully, in many instances) as the result of an event versus planning for and building to withstand events. The classic example here is planning for 1 in 100 vs 1 in 1,000 or 10,000 year events. The Dutch are well known for this relative to their sea level exposure and the robustness of their defense and resiliency systems to avoid catastrophic damage. Their approach has been contrasted, rather dramatically, with the poor resilience preparation of many American coastal cities, including those in southern Florida and the Gulf Coast which are severely threatened by sea level rise and hurricanes. Additionally, from the earth systems perspective, the increased fidelity of climate scenario modeling, and the ability to downscale those into extreme potential events and their impacts, is also adding to the ability of resiliency planning.

As introduced in IPCC AR5 and further elucidated in AR6, the idea of "climate-resilient" energy systems, and an energy economy, is even broader than first conceived. Here, the IPCC has introduced the concept of mutually resilient mitigation and adaptation planning, which might be considered a different language compared to the concept that was just described earlier. However, it is gaining additional attraction at the highest government levels due to the increased attention for the need for climate adaptation, as well as the speed and scale of decarbonization and mitigation measures.

Clean Electricity is the Backbone of Our Clean-Energy Economy

The view from the human perspective of connectivity, comfort, goods, and mobility complements the standard energy segmentation view. Here, traditionally, energy sector consumption has been viewed as residential, commercial, industry, and transportation. Energy supply was separated into primary sources such as wind, solar, geothermal, hydropower, coal, nuclear, petroleum, natural gas, and biomass, and then intermediates such as electricity and hydrogen. Intermediate or prime energy supply was then mapped to the consumption sectors. In a net-zero economy, some of this map stays consistent. Other elements either disappear or are complemented by carbon capture utilization and storage plus carbon removal technologies, or from a broader climate perspective, climate management actions, such as afforestation or geoengineering. With significant transition of the transportation sector to electrification and other clean fuels, petroleum's role in a net-zero economy is anticipated to be significantly smaller. As noted, on an economic basis, going forward, renewable electricity generation will dominate new capacity build-out for the decades to come. However, there will remain some fossil fuel generation if combined with carbon capture. Let's unpack that a bit, starting with the power system.

Realizing a zero-carbon power system

Building on the transformational work of the Renewable Electricity Futures Study and the body of knowledge that was built in the previous decade, it is readily apparent that renewables, in particular variable wind and solar, can provide a substantial amount of affordable, reliable, and secure power, when combined with advances in forecasting, operations, demand management, and appropriate-duration storage systems. One might ask if a power system can run only on 100% renewables. The question of 100% is not technical, but economic and highly dependent on the location. That is, a 100% renewable system may include hydropower, biopower, geothermal, solar thermal power (with storage), wind and solar PV (likely paired with batteries), potentially ocean energy, and other renewable resources. Some combination of those may allow a system to be designed and operated with 100% renewable assets. However, not all renewable generation is created equal or equally economic. They are location specific. Geothermal energy and hydropower are location dependent. Wind and solar generation varies dramatically by location as well. Availability of large areas for geospatial diversity opens up greater potential but may also imply difficult-to-build transmission or access to remote locations which are more expensive. But, the point here is not to argue for 100% renewable systems; it is to point out that renewable systems at high penetration are feasible, viable, and reliable, and that system definitions might actually be broader than that. Stakeholders may choose to have net-zero carbon in the power system or "100% clean energy." That categorization, of course, is ambiguous and up to the given stakeholders. It might also be that some emissions would continue to be allowed in the power sector and offset by negative emissions and other sectors, such as through land use and carbon dioxide removal technology such as direct air capture (DAC). Each of these trade-offs will play out at a local, regional, and national level depending upon the resource base, the economics, and of course the socio-political and institutional factors.

Given the projected growth and continuing economic attractiveness of renewables, it is worth diving into renewable-based power systems and energy systems a bit further. Multiple studies have shown that it is readily straightforward to incorporate 80–90% renewables, with 50–70% of those being variable. Making up the last 10–20% of that supply with low-carbon resources is a trade-off between building sufficient resources for resiliency and long-duration storage for times when weather conditions limit real-time generation to a level that would not be sufficient to meet the load, using biomass resources which may be categorized as renewable or anthropogenic fossil resources with carbon capture and storage, and, of course, using nuclear. Jurisdictional goals vary. Some of them are set as 100% renewable. Others are set as 100% clean. Others are set as net zero. Austria, for example, committed to a zero-carbon electricity system by 2040.[98] The U.S. set a goal for zero-carbon electricity by 2035.[99] Numerous other goals have been announced by countries, counties, cities, or companies. The technical solutions to each of those are different. What is common among them, as multiple studies have shown in America, Europe, and elsewhere, is that because renewables are the least cost generation options today and going forward, they will be the bulk of new power generation (they were nearly 90% of all new generation capacity added in 2022 globally). Further, renewables will become the largest source of overall power capacity and generation going forward.

It is worthwhile reflecting upon a few of those studies. Using the U.S. as an example, there are four studies that offer valuable insights, both comparable and contrasting. The four studies are the Net-Zero America Study by Princeton University,[62] the analysis of the 50–52% reduction goal of 2030 by the Rhodium Group,[100] the analysis of 100% "clean" electricity by 2035 by a team at NREL,[101] and the analysis of net zero by 2035 by the Electric Power Research Institute (EPRI).[102] Reflecting across all of these studies, the various methodologies that they use and the data that are used offer some common perspectives. First, achieving the goal set out for 100% decarbonized electricity by 2035 would imply significant increases in investment and low-carbon technologies as well as retirement of existing high-emitting technology such as unabated coal plants. Second, the combination of technologies used to achieve the decarbonization goal of 2035 is highly dependent on the cost trajectories of those technologies as well as the interpretation of the term "decarbonization" and whether or not offsets are allowed in other sectors. For example, the EPRI Study evaluates the options for bioenergy with carbon capture and storage in the power sector and contrasts that with a carbon-free target that includes nuclear but not CCS and a 100% renewable scenario that includes hydrogen as a combustion fuel and battery storage. The NREL Study evaluates numerous scenarios with various technology combinations including new CCS, or excluding it, or potentially extending nuclear power beyond its current licensing time frames. The Princeton Study evaluates a similar set of technology options and further looks at the infrastructure build-up requirements needed to reach those goals. Overall, there is consistency among the studies which shows that achieving 80–90% decarbonization by that time frame is relatively straightforward using today's commercial technologies. The remaining 10–20%, however – as a function of a set of technology assumptions, cost assumptions, and definitions – is critical to ensuring the reliability of the power system in 2035 and beyond, and moreover affects the overall cost structure. Agnostic of the specific technology mix for decarbonization of 80–90% of power generation, the implications are that the technologies for the last 10–20%, if applied within the power sector only, are the target of needed research, development, and scaling to reduce the costs and improve the technology performance and competitiveness. This includes batteries and long-duration energy storage including power to hydrogen or other gases, or power to liquids. The solution set expands as one considers broader economy-wide decarbonization, in particular the transportation or mobility sector. Given recent reductions in power sector emissions in the U.S., and the persistence of fossil fuel use for cars, trucks, buses, airplanes, and ships, the transportation sector is the major contributor to emissions in the U.S. and overall economies around the world. The implication of this situation is that the continued evolution of low-carbon technologies, including electric vehicles for light-, medium-, and heavy-duty usage, SAFs, and low-carbon, electrified mass transit, will be significant. What is consistent across all the scenarios is that energy efficiency and the continued expansion of "smart, efficient energy management" are fundamental to reducing the investment requirement and overall energy generation needs to meet future energy demands. These overall conclusions are robustly consistent with the findings of analyses that are being conducted for economies around the world.[103] The importance of energy efficiency, in particular the provision of highly efficient buildings and the evolution of low-carbon mobility solutions in developing economies, cannot be understated. This is because the growth of the energy demand associated with increased building stock and increased mobility, including both ground and air transport, represents significant future projected energy demands and, if not abated, greenhouse gas emissions.

Let us turn now to additional efforts that are helping translate ambitions into implementation plans. In numerous locations both corporate and jurisdictional leaders have committed to decarbonization goals. The key question is how one can achieve those goals in a reasonable time frame and what the implications are for infrastructure build-out, cost, and program development. Let's look at a leadership example. In 2016, the then mayor of Los Angeles, Eric Garcetti, set a bold ambition for the city to be 100% renewable by 2045.[104] To reach these goals and assess the implications for jobs, electricity rates, the environment, and environmental justice, the Los Angeles City Council passed a series of motions in 2016 and 2017 directing the Los Angeles Department of Water and Power (LADWP) to determine the technical feasibility and investment pathways for a 100% renewable energy portfolio standard. The LADWP is a municipal utility owned and operated by the city. The LADWP reached out to the National Renewable Energy Laboratory to lead the study of the implications for this bold ambition, as the laboratory has unparalleled depth and breadth for conducting planning studies to inform capital investments and programs. Conducting that study involved monthly meetings with community stakeholders and LADWP staff. Stakeholders had a wide range of concerns and questions. They were very concerned about equity, energy, and environmental justice in this transition. They were also concerned about what energy sources could be included. They asked for an evaluation of options that included, or excluded, natural gas, biopower, and nuclear power. The group eventually settled on four scenarios, one of which included an earlier target date of 2035, with no biofuels or natural gas and the use of existing nuclear facilities. Other scenarios evaluated the 2045 target with various combinations of either new transmission, allowing (or not) natural gas or nuclear, and different levels of electrification of building loads and transportation. The city utility owns and operates assets not only in Los Angeles County but also throughout California and in neighboring states. The geographic diversity and the county's openness to developing geographically dispersed resources were also factors. The stakeholders were interested in maximizing the amount of in-basin generation, that is, generation that could comply with their definitions and be within Los Angeles County. They were also interested in further electrifying the demands within the county, including residential demand such as heating, cooking, cooling, water heating, as well as the expansion of electric vehicles for both household use and school buses. Taking all of these considerations into account, a complex set of models was used, leveraging more than 50 TB of data and hundreds of millions of simulations on a supercomputer. The modeling of every building in Los Angeles County, more than 7 million of them, involved 3.6 million processor hours, or more than 60 years of processing on a laptop. Satellite imagery was used to identify appropriate rooftops and other locations for distributed solar power. Traffic pattern data were used to project potential electric vehicle use and charging requirements. All of this was done at the block level and down to five-minute simulation increments. Further, reliability was a critical concern. The simulations and scenarios had to be run in a manner that reflected the potential loss of critical transmission lines or of multiple generation facilities that were simultaneously unavailable. Results indicated that variable wind and solar energy could supply 70–90% of the final energy demand. It is interesting to note this is on the same potential scale as the International Energy Agency's findings for net zero by 2050 for the world. Other results showed the importance of utilizing renewable power to potentially create hydrogen or renewable natural gas. Both of those chemical fuels were found to be critical for reliability of the system. Another interesting result was that the city could support the build-out of gigawatts of distributed solar energy with only minor changes to their distribution system upgrade plans. Under all the scenarios, significant air quality and health benefits would be realized with a significant portion of those in the underserved communities. Economically, while the transition to 100% renewable would be more expensive than business as usual or the state plan for 100% clean electricity, the overall economic cost was relatively low compared to the economy of Los Angeles County. The study was released in March 2021. Shortly thereafter, the city council adopted a more aggressive goal than the mayor had initially championed. That is, the council directed the LADWP to achieve 100% renewable by 2035. This was a leading commitment not only in the U.S. but also elsewhere for a city of that size. Since then, the LADWP has begun implementation, including advancing its hydrogen strategy, as well as engaging in deeper equity strategy studies to ensure that clean-energy transition is inclusive and as fair as possible for all residents of the city.[105]

The complexity of the modeling, including the technical, spatial, and temporal resolution, was unprecedented. The "model map" is shown in Figure 2. The left side of the diagram is focused on load development, in which the end uses (cooking, heating, and lighting) were modeled at the building level and then aggregated for

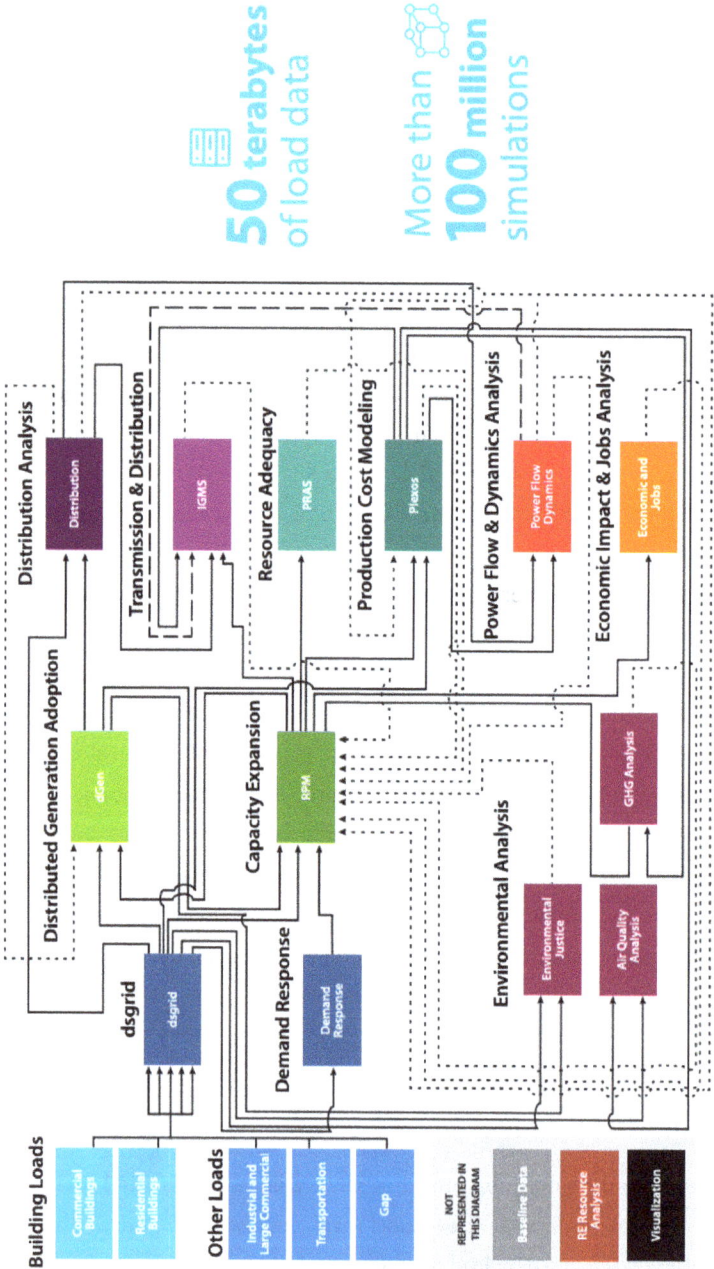

50 terabytes of load data

More than 100 million simulations

Building Loads
- Commercial Buildings
- Residential Buildings

Other Loads
- Industrial and Large Commercial
- Transportation
- Gap

NOT REPRESENTED IN THIS DIAGRAM
- Baseline Data
- RE Resource Analysis
- Visualization

Distribution Analysis
- Distribution

Transmission & Distribution
- IGMS

Resource Adequacy
- PRAS

Production Cost Modeling
- Plexos

Power Flow & Dynamics Analysis
- Power Flow Dynamics

Economic Impact & Jobs Analysis
- Economic and Jobs

Distributed Generation Adoption
- dGen

Capacity Expansion
- RPM

Environmental Analysis
- Environmental Justice
- Air Quality Analysis
- GHG Analysis

dsgrid
- dsgrid

Demand Response
- Demand Response

Figure 2.   Map of modeling interactions and data flows for LA100 study.
Source: Reprinted with permission from NREL.

local (distribution level) loads and eventually system-wide loads. Electric mobility charging was modeled by using anticipated population, traffic patterns, EV fleet size, and characteristics (e.g., battery size) and an assumed charging preference (e.g., daily or every other day).[106] Estimates of industrial and commercial demand response were built up based on prior technical analytic approaches and applied to the portfolio of companies in the LADWP service territory. Distributed generation, predominantly by PV and sometimes in combination with batteries, was evaluated based on satellite imagery for rooftop space and location, and included other applications such as parking lots, potentially degraded lands, and even small "floating-PV" applications (PV systems on water bodies such as ponds). Classic capacity expansion modeling was done using a model that captured details both within the LADWP service territory and larger regional interconnections. This type of modeling is sophisticated economic optimization which matches generation and load, accounting for reliability requirements as well as spatial and temporal characteristics of the power system. The right side of the model map identifies a number of more detailed analytics that were used to evaluate system reliability scenarios including major transmission outages for potentially months at a time and then also identify distribution system needs. Detailed operational modeling, "production cost modeling," was also performed at very high spatial and temporal resolutions across the whole system in order to provide insights to the engineering staff that the system would indeed operate reliably given the fairly dramatic changes in the load and generation portfolios. Additional modeling included air quality modeling, greenhouse gas modeling, and economic modeling.

LA100 is just one example of the application of the modeling advances that had progressed in Phase III. From a reliability and system operations perspective, the innovations noted in Phase III in which inverter-based resources are now able to supply reliability services, such as ramping their power up and down, or supporting the frequency and inertia of the system, have opened up new opportunities and perspectives for system operators around the world. Additionally, leading system operators are learning how to incorporate significant amounts of distributed energy resources, in particular solar photovoltaics. Here, Australia is leading the way with nearly 80% of households in South Australia having rooftop PV systems. Current innovations are focused on increased visibility of the potential power output of those systems and controllability. As in California, any system with a significant

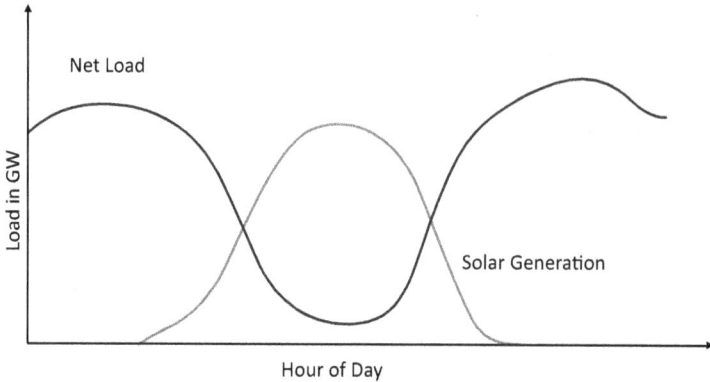

Figure 3.   Duck curve showing daytime PV production and its impact on net load.

amount of solar photovoltaics on it will experience the daily generation profile that is now fondly known as the "duck" curve.[107] A conceptual diagram of a duck curve is shown in Figure 3. Managing the evening ramp up in demand just as the sun is setting has proven to be a persistent challenge. The introduction of short-duration batteries to shift some of the daytime peak solar output to the evening ramp is one solution. Other solutions include bringing on gas-fired turbines, other storage solutions such as pumped hydro or mechanical storage, and of course evening load growth management.

Integrated systems: Tightly coupled hybrids

The energy system of the future is not the same as that of the past. Previously, the evolution that is taking place was described as follows: a transition from large centralized plants with one-way power flow through the transmission and distribution networks to serve loads to a system of increasingly digitized, decentralized, distributed networks, in which energy supply may be significantly provided at the same location as the end use; all assets which are associated with the grid may be digitally addressable to more actively be a part of real-time balance of supply and demand. Within this increasingly heterogeneous decarbonized system of the future, there are additional configurations that will also be developed. Some of these additional configurations fall within the concept of integrated, tightly coupled hybrid systems. What does that mean? Let's consider an operating example today. In the early 2020s in Morocco, the government was asking for bids to provide desalinated water at a given location. The assumption going in was that the project developers would essentially design a desalination system using reverse osmosis or other technologies and buy all of the power from the existing grid. An alternative approach was proposed in which the project developer combined on-site solar generation with the desalination. The on-site solar generation reduced the electricity system requirements and ongoing operating cost, and provided power to the desalination plant at a price cheaper than purchasing the electric system power. Nearly a dozen hybrid projects with renewables and water desalination are commercially operating today, mostly throughout the Middle East.

Another relatively simple example involves the use of water cooling for photovoltaic panels. Photovoltaic panels perform better when cooler. Here, the integrated system not only improves the performance of the PV array but also provides hot water that can be utilized for multiple applications.

The concept extends even further. Consider a tightly coupled system in which renewable generators are coupled with a nuclear power plant. The combined set of generation assets is configured to be able to provide power, purified water, hydrogen, or heat. System designs and operating parameters are created in order to minimize costs and maximize revenues across the various product streams. Each of these can be effectively forecasted and served in real time or via integrated storage at the tightly coupled hybrid system facility.

Interestingly, another tightly coupled innovation was proposed a number of years ago that combined a nuclear power plant and energy storage. In this configuration, the nuclear plant would be able to run in a relatively constant operating mode and the thermal storage would provide a buffer between the nuclear plant and the power grid. That buffer would enable the combined system to be effectively dispatchable and provide power only when needed by the system operator. Today, multiple next-generation nuclear companies are designing these hybrid systems using molten salt as the thermal storage technology. Whether or not they will succeed at a commercial scale in the near term will depend upon regulatory approvals and economics. Alternatively, there are a number of nuclear fusion start-ups. Fusion is the process in which deuterium atoms are fused together in a highly exothermic reaction that creates more energy than the input energy. These types of reactors are operationally safe by design and based on fundamental physics, since they hold no nuclear radioactive materials and immediately shut down once the input power is cut off. Most recently, in April 2023, the U.S. Nuclear Regulatory Commission indicated that these plants would be regulated under a completely different set of rules compared to the fission reactors that have been operating for the past many decades. A few of the fusion start-ups have announced plans for demonstration projects in the next three years and are hopeful for commercial-scale deployments in the early 2030s. The world will have to wait to see.

Why would one create such a complicated hybrid system? Early analytic studies have indicated that tight coupling offers economic opportunity to arbitrage across the product streams while utilizing the availability of the generating assets to their maximum.[108] Independently, those assets would be competing in a marketplace that is likely limited only to power. The other applications are water processing or the creation of hydrogen or heat-independent contracts or services, which limit the financial and operating flexibility. This multi-input, multi-output configuration contrasts with a single-input, multi-output configuration. An example of the latter is the Diablo Canyon nuclear generation facility in California for which analysis indicated that extending the life of the plant, and converting it to a poly-generation facility with the option to produce power, clean water, and clean hydrogen, could offer billions of dollars of savings to California ratepayers and help relieve some of the water stress in Southern California as well as initiate a clean hydrogen industry in the state.[109]

Power to X (For fuels, chemicals/materials, and industrial processes)

Here, the story extends from zero- or low-carbon power systems to the rest of the energy economy. When one conceives of a zero-carbon energy economy, one has to consider all of the uses of energy throughout the energy economy. Classically, one might consider a Sankey diagram of energy flows.[110] Here, the primary energy sources and resources are displayed on the left as "inputs." They are tracked through their conversion processes to different end use sectors such as in buildings, transportation, and industry. The flows articulate interconnectivity among our energy sectors. For example, oil and natural gas are inputs not only for chemicals and fuels for industry but also for power. Nuclear power provides energy input to buildings, mobility, and industrial sectors. Renewable power is similar to nuclear power; biofuels flow into mobility/transportation and direct renewable heat flows into both buildings and industry. The new intersection among the sectors across the economy is clean hydrogen. Combining clean hydrogen with non-primary anthropogenic carbon molecules (from biomass or carbon capture or DAC) opens up a whole new set of chemistries for creating low- or zero-carbon fuels and chemicals.

Contrasting two Sankey diagrams is a useful exercise. Figure 4 (a) shows the Sankey diagram for world energy for 2021 from the International Energy Agency.[111] On the left-hand side, one sees the relative contributions of the primary energy sources in EJ: oil, gas, nuclear, coal, renewables, etc. Various conversion pathways are shown and then the end uses are provided on the right-hand side. Note the fairly significant amount of losses in the conversion of many of those fossil-based resources to useful energy. This comes back to the efficiency of those burning processes, such as coal power plants. In 2021, use of significant fossil energy produced approximately 40 Gt of greenhouse gas emissions worldwide including exploration, production, transportation, and leaks.

In stark contrast to 2021, Figure 4 (b) represents the conceptual Sankey diagram of a net-zero-energy world in 2050 according to the models of the International Energy Agency. This is remarkably different as one can see from looking at the diagram. Fossil resources are significantly reduced, and there is dramatic expansion of clean electricity in stark contrast to 2021, even with all the successes to date. One might recall that the IEA anticipates that electricity will supply on the order of 60% of end use energy needs by 2050, and 60–70% of that electricity will come from wind and solar power alone, but there is also significant growth in new next-generation nuclear, hydro, biopower, solar thermal, and geothermal sources (other renewables in Figure 4 (b)). It also shows new conversion processes across sectors. This is where advanced electrification comes into play for industrial processes, but also where the conversion of electricity to hydrogen or other chemicals and fuels becomes a critical interdependency across sectors. Lastly, according to the modeling, and consistent with the analysis of the IPCC and others, some fossil energy will remain critical to the production of chemicals and fuels and some industrial processes. In a net-zero world, GHG emissions associated with these uses are reduced by CCUS and offset by DAC or other carbon removal processes, including bioenergy with CCUS (BECCS), land use management, and other strategies.

Let us look at some of these conversion pathways in more detail.

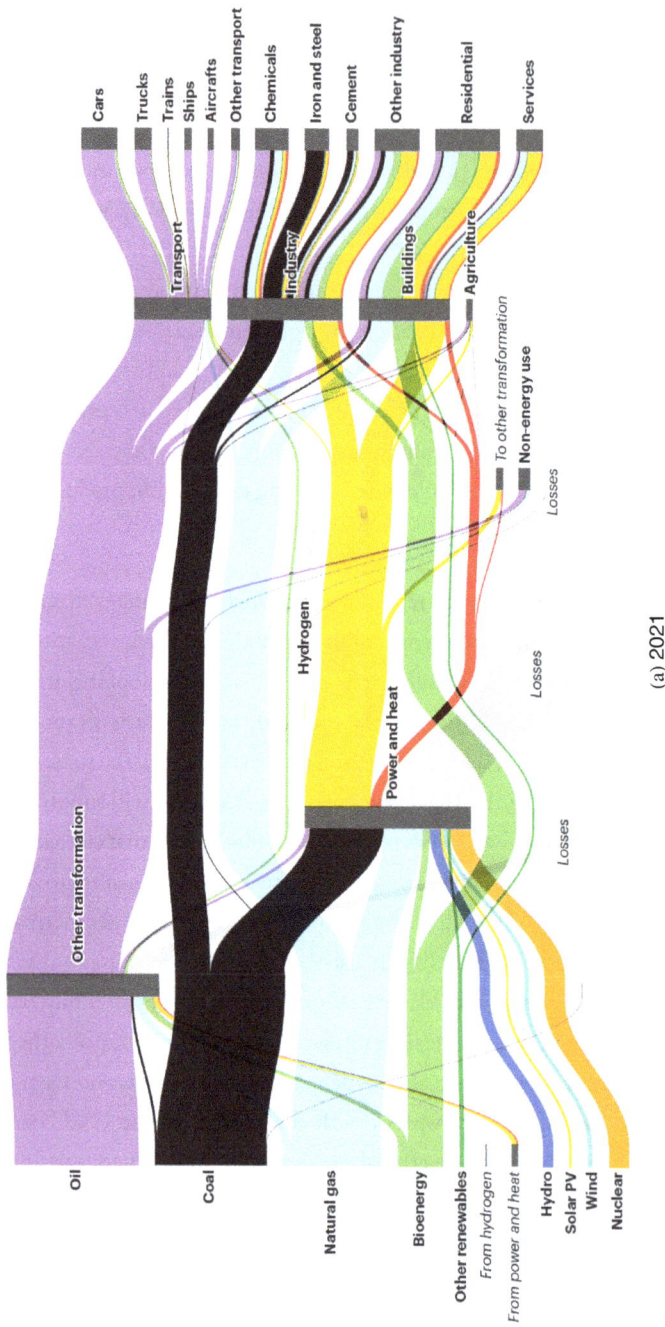

Figure 4. (a) Sankey diagram for world energy flows for 2021. (b) Projected world energy flows under Net-Zero Energy Scenario in 2050.
Notes: Some electricity is used to generate hydrogen from water electrolysis, while some hydrogen (and hydrogen-based fuels such as ammonia) is in turn used for power generation in 2050. Losses include fuel, heat and power distribution losses, as well as transformation process conversion losses and own use. Electricity becomes the largest energy vector in the NZE Scenario, with demand more than doubling between 2021 and 2050.
Source: Reprinted with permission from IEA.

(a) 2021

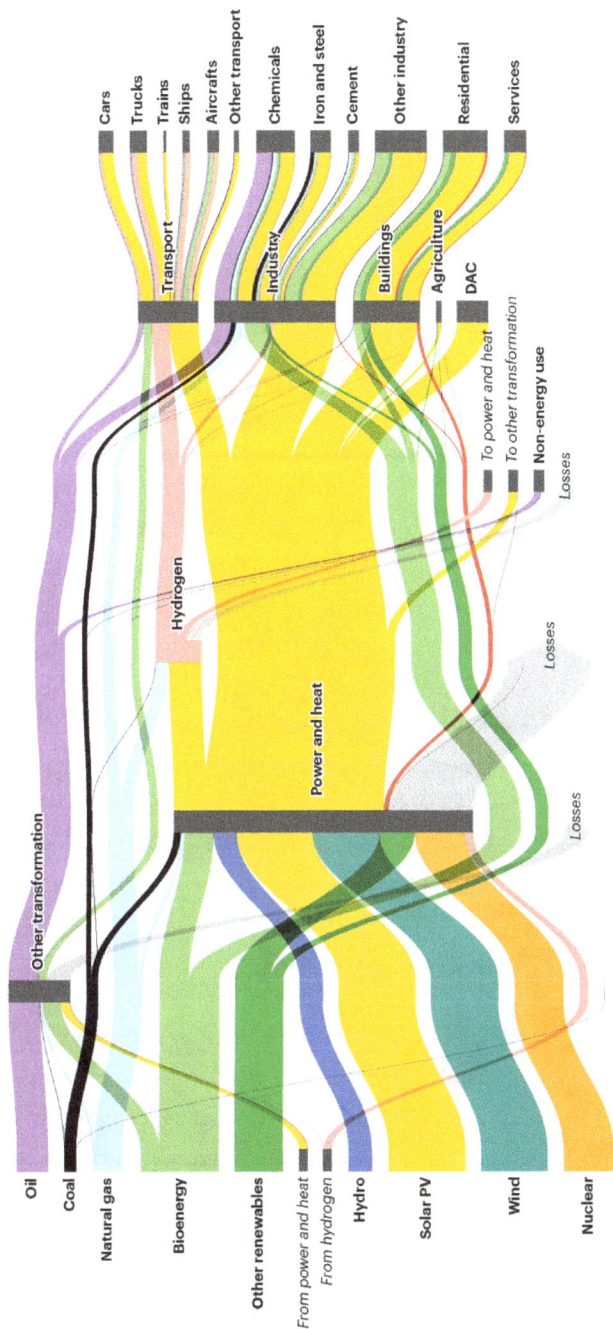

Cars
Trucks
Trains
Ships
Aircrafts
Other transport
Chemicals
Iron and steel
Cement
Other industry
Residential
Services

Transport
Industry
Buildings
Agriculture
DAC

To power and heat
To other transformation
Non-energy use
Losses

Hydrogen

Power and heat

Losses
Losses

Other transformation

Oil
Coal
Natural gas
Bioenergy
Other renewables
From power and heat
From hydrogen
Hydro
Solar PV
Wind
Nuclear

(b) 2050

Figure 4. *(Continued)*

Hydrogen

Hydrogen (H_2) is one of the simplest molecules. It has an incredibly diverse array of uses for energy and chemicals. For example, it is burned with oxygen to propel rockets. It can be combined with nitrogen to create ammonia. It is a core "workhorse" for combining with hydrocarbons to create gasoline, jet fuel, and other chemicals from crude oil to natural gas or biomass. It can be made via an electrolyzer. H_2 can be converted to electricity directly via a fuel cell, or blended with natural gas or burned in a combustion turbine with air or oxygen. It can be used to directly reduce iron oxide to produce steel.[112] This diversity of uses and the recent, and growing, cost competitiveness of producing low-carbon H_2 from either clean electricity or via CCUS has garnered unprecedented attention and investment.[113]

Given all that, the question is why hydrogen has not been used within those diverse processes and end uses at scale before. The issue resides in supply as well as cost competitiveness within those processes. Global hydrogen production in 2021 was 94 MT/year and is anticipated to grow toward 200 MT/year by 2050 under low-carbon scenarios. It is predominantly, more than 98%, produced via cracking natural gas that produces significant GHG emissions (~1,000 $MtCO_2eq$). To understand that, consider steam methane reforming, which converts $CH_4 + H_2O \rightarrow CO + 3H_2$. CO is carbon monoxide, formed by the primary reaction under heat and in a catalytic oxygenated environment. That is, for every three molecules of hydrogen produced, one molecule of CO is produced. CO is then further reacted with steam to produce CO_2 and more H_2; $CO + H_2O \rightarrow CO_2 + H_2$. Chemically, about 5 kg of CO_2 is produced for every kg of H_2. When one includes the CO_2 for heat, it increases to 9 $kgCO_2/kgH_2$.[114]

In contrast, consider $2H_2O \rightarrow 2H_2 + O_2$. That is, water is "split" by electrochemical mechanisms to produce hydrogen and oxygen. Both pure molecules can be readily separated and used for different applications. However, water splitting requires both clean electricity and highly efficient, low-cost electrolyzers, particularly to economically compete with natural-gas-based generation without incentives (or penalties) for reducing carbon emissions. Scale is also a factor. For example, globally nearly 40 Mt of hydrogen is used in refining processes annually. Production of that quantity of H_2 via electrolysis would require approximately 600 GW of power and electrolyzers (this is highly dependent on the efficiency of the electrolyzer). Electrolyzers on the GW scale have not been commercially available until very recently, nor cost competitive. The efficiency (and cost) of the electrolyzer is key to cost-effective hydrogen production. Measured in kg/kWh, the efficiencies of most commercially available electrolyzers (either polymer electrolyte membrane, PEM, or alkaline) ranged around 75%, implying a fairly significant loss on an energy basis. In 2022, it was announced that up to 95% efficiency was promised, which would continue to drive costs down per unit of hydrogen production.[115]

A groundbreaking example of the new green hydrogen opportunities was the announcement of a $5B NEOM green hydrogen/ammonia facility announced by a partnership of Acwa Power and Air Products. NEOM, incorporated in August 2021, will produce up to 600 tons per day of carbon-free hydrogen in the form of green ammonia as a cost-effective solution for transportation globally, based on 4 GW of solar and wind generation.[116]

Hydrogen seems to be the "darling molecule" for the next decades and centuries. Countries around the world are establishing hydrogen strategies, in particular green hydrogen strategies. They recognize the synergies and opportunity for co-development of renewable power and the use of that power to create hydrogen and deliver zero-carbon hydrogen-based energy around the world. The opportunity space as outlined in the IEA's net-zero scenario is actually quite significant, particularly given hydrogen's role or derivative molecules from clean hydrogen in a net-zero energy economy. It is also viewed as a new export opportunity leveraging renewable domestic resources.[117] Interestingly, the lure and attractiveness of green hydrogen is pervasive throughout the energy community as well as the finance community. While few projects are actually reaching financial closure with attractive returns, the prospects are enticing to many, with the number of such ventures increasing. A recent report by the International Monetary Fund includes their perspective: "Clean hydrogen has the potential to upend the geopolitics of energy as we know it."[118]

From a technical viewpoint, as noted earlier, the chemistry is relatively straightforward. The engineering and cost economics, however, are not. Further scaling of electrolyzers as well as fuel cells or other processes may represent significant barriers to achieving this hydrogen vision. Globally, competing with green hydrogen and its derivatives implies concomitant development of clean electricity systems at a much larger scale. Realizing power systems that could be 3–10 times larger than those today represents both an enormous opportunity for growth and an enormous set of challenges. Hydrogen plants may or may not be connected to the main grid depending upon how they are configured and operated, based principally on the economics. That is, there could be dedicated off-grid wind or solar to hydrogen to ammonia plants. Alternatively, those plants may be connected to the grid with arbitrage between selling power, taking power from the grid, and producing hydrogen. The hydrogen might be used directly or converted to ammonia, methanol, or other chemicals and fuels. This is all yet to play out and will depend on how the economics and infrastructure develop in different environments and different countries, as well as the geopolitical dynamics that will develop with the expansion of the global green-hydrogen-based chemical trade.

Ammonia

The NEOM project is a prime example of conversion of green hydrogen to green ammonia, which will then be shipped worldwide for reuse. Ammonia is the foundation for fertilizer, and approximately 150 Mt/year is produced via commercial (Haber Bosch) processes, with "grey" hydrogen derived from natural gas. Ammonia can be a "transport molecule" for hydrogen, or it can also be used as a chemical input or as a fuel. An extensive review of nitrogen-based fuels was published a few years ago.[119] In that review, multiple different production pathways were outlined, as well as the relative comparisons of direct end uses, including as a hydrogen carrier. On the production side, there continue to be interesting innovations which are not yet commercial, but open up intriguing opportunities. For example, a complex multi-catalyst membrane configuration has been created to enable the co-production of ammonia on one side and CO_2 and hydrogen on the other side based upon input of CH_4 and air (N_2, O_2, and H_2O). This multistep heterogeneous configuration could potentially reduce energy requirements by more than 50%. However, one would have to combine it with captured CO_2 to reduce the GHG footprints, or replace natural gas with renewable-based methane.

As a carrier for hydrogen, utilizing lithium hydrate as a catalyst for ammonia decomposition offers commercially scalable and relatively low-cost opportunities for reconversion to hydrogen. Lithium hydride catalysis occurs at relatively low temperatures, below 300°C, which is readily obtainable through renewable energy heat generation or other low-temperature production (or byproduct) technologies. Other pathways, including hydrazine and borane, have also been considered but are less developed and not available commercially today.

Molecular foundries: Low or zero anthropogenic carbon fuels and chemicals

Here, the landscape is much, much wider. Considering the breadth of chemicals used in our energy (and materials) ecosystems, that use could be derived from biomass or captured CO_2, H_2, and N_2. Many of the core chemicals that form the basis for industrial production, materials synthesis, plastics, fabrics, etc., have been derived from petrochemical inputs via a complex suite of relatively efficient, high-volume processes – regretfully many of which emit large amounts of CO_2 or other GHGs. In order to reconsider this ecosystem as a zero- or low-carbon ecosystem, there are two fundamental options: one, reducing the energy and GHG intensity of the existing system through electrification, adding CCUS, and potentially developing alternative reactant supply chains; two, replacing natural gas and syngas with non-anthropogenic substitutes via utilization of biomass or captured CO_2 and renewable or zero-carbon H_2 as principle building blocks to "build up" the desired chemical. This contrasts starkly with traditional processes that start with heavier hydrocarbons and break those down via catalysts, temperature, and pressure and then isolate pure streams. Creating a low- or zero-carbon chemical and fuel industry would naturally leverage the existing knowledge base and infrastructure and exploit "catalytic" innovations to avoid or eliminate GHGs, including those associated with process manufacturing (heat production, compression/pressure, etc.), as well as primary reactants. However, the bulk of our current chemical and fuel production is based on traditional petroleum-refining processes. Reconfiguring these plants to become molecular foundries would involve considerable investment, process innovation, reengineering, clean power investment, and clean hydrogen production, including the development of pure CO_2 reactive streams either from biomass or from CCUS. While this represents enormous opportunity for new innovation, it also represents enormous financial risk to large, established industrial companies. The competitive landscape for providing the required chemicals and fuels is, of course, not a singular substitution dynamic. As noted, on the fuel side, there are significant competitive pathways, including electricity, hydrogen, ammonia, and methanol. On the chemical side, there is an increase in pressure to reduce waste, increase recycling, and develop circular pathways in order to produce the demanded primary inputs, and reduce the associated GHGs and other environmental impacts.

Material circularity: Design for recycling, reusing, and repurposing

Material circularity has matured in concept from the early days of simply recycling. Today, it is much more comprehensively considered as a holistic approach to design for reusing, repurposing, remanufacturing (into potentially other chemicals), and recycling. The approach to this includes multiple different factors, such as developing novel materials, creative designing, architecture to increase recyclability, designing manufacturing processes that reduce material and energy use, increasing technology lifetimes, and advancing social, economic, and regulatory decision environments to support advanced circularity.

In the energy economy, the principal reuse, recycle, and repurpose activities are advancing rapidly for batteries, PV, and plastics, with additional elements focused on building materials, polymers, and composites that are used in wind turbines, and of course metals that are used throughout the power system. New approaches to materials include comprehensive, full-lifecycle assessments, not only on the material side but also on the energy side. For example, in buildings, new material approaches involve comprehensive physical and spatial modeling of material energy envelopes, including new methods of modular manufacturing and reduction of site waste.[120] These approaches extend beyond and are complementary to the Sustainable Forestry Initiative, for example, which focuses on sustainable forestry practices for wood production, but does not address the rest of the lifecycle materially or energetically.[121] For PV, there is increased public–private partnership activity to advance the practices of reuse and recycle. The industry, non-governmental organizations, and governments all see considerable need for advancing the solutions, with the increased annual installation of PV heading toward 500 GW and potentially reaching the TW/year level by 2030. Even for approaches that would extend panel lifetimes to 50 years, this annual installation growth would imply multiple tens of TW panels to be reused, reproduced, or recycled in the late 21st century.[122] Similar levels of ambition, collaboration, and investment are gaining momentum in the battery sector as well.[123]

On the plastics side, considerable news has been announced over the past few years. Extensive plastic waste has been generated, and it is now prolific in the world's ecosystems. This waste, of course, is estimated at hundreds of gigatons of microplastic particles in the ocean which are affecting all aspects of the ecosystem including mankind. To give a sense of this, nearly 6,000,000 metric tons of discarded plastics per year are not recycled. Recently, there have been considerable technical innovations developed which hold promise for what can be called single-stream recycling. In this approach, creatively designed molecules are able to decompose multiple different plastics into fundamental building blocks in order to build a principle set of chemicals for a supply chain for recycled plastics. Multiple pathways are under development that range from thermal catalysis to electro catalysis, photo catalysis, and combined activation mechanisms (e.g., photoelectro or photothermal). Each of these particles can be deconstructed into multiple diverse parts and formed into fundamental building blocks and then upcycled or repurposed as recyclable polymers.[124]

From a governmental perspective, one can see cases being announced around the world. For example, in 2022, the European Union made an early declaration called the new Circular Economy Action Plan (CEAP), which is part of the European Green Deal.[125] The CEAP encompasses a strategic roadmap as well as core funding and regulatory and control processes to advance sustainable products as the norm in the European Union, and focus on sectors that use the most resources such as electronics, batteries and vehicles, packaging, plastics, textiles, construction, food, and water. Major public–private partnership initiatives, such as those from the World Economic Forum Platform for a Circular Economy (PACE), have formulated bold visions and missions to "catalyze global leadership from business, government, and civil society to accelerate the transition from a linear to a circular economy that will improve human and environmental well-being for current and future generations."[126] All of this holds promise but has yet to materially deliver economically viable and environmentally sound solutions at scale, in particular in Africa, Southeast Asia, and other emerging economies.

Economics/Finance: From Competitive to Mainstream

In prior Phases, technology's cost competitiveness was discussed from an economic and finance perspective, as well as the perspective of financial engineering and the capital financial markets. Here, those factors will be discussed, in addition to a broader range of factors that now need critical thinking in terms of how the world can transition to and prosper in a net-zero economy.

The transition from today's extractive and high-GHG economies, in most places dominated by fossil fuel extraction, to low- or zero-carbon economies is most critical to navigate. Trillions of dollars of infrastructure are in place and operating. Trillions of dollars of investment are at risk. It takes a "steady hand" and a real long-term strategy to navigate the turbulence of the transition. Here, there are great differences in the energy transition viewpoints among sovereign wealth funds and state-owned enterprises versus proactive multinational oil and energy companies, and older, entrenched businesses that have been running for a century or longer, such as the coal mining and power plants that populate the world. My colleague Jason Bordoff, now at Columbia University, has written quite extensively on this within the geopolitical realm. Some of the work of Bordoff and his colleagues can provide some guiding insights into the careful navigation that is required over the next multiple decades.[127] Today's infrastructure stands in stark contrast to that which needs to be developed and operated in the next 15–30 years to build renewable energy technologies at a massive scale including other low-carbon solutions when effective and financially viable, such as carbon capture, utilization, and storage, and even new nuclear power in locations where it is permissible and would be cost competitive. Policy incentives such as the IRA bill in the U.S. potentially have a profound effect on this portfolio of technology options and the pace and scale of achieving decarbonization.

What perhaps is needed for this industrial transformation is a fundamental rethinking of the economic development pathways in the world, and the role energy provides in those, as well as much more complex systems-level thinking, understanding, planning, and implementation. Here, it is important to reflect on the interconnectivity among sectors: for example, across the energy–food security nexus. At this nexus, broadly, the importance of fundamental systems-level thinking is paramount. It takes energy to pump, purify, and distribute water among crops, run tractors, create fertilizers, run processing facilities, and transport crops and other products. The connection with security is a conflict as old as man regarding access to resources for living, economic prosperity, and thriving. This has become even more complicated over the last century with the intense complexities of geopolitics, resources, geopolitical positioning, and infrastructure to provide for safe, reliable, and economically viable pathways for development and economic prosperity. This broad perspective, while perhaps recognized qualitatively by a few, but not all, contrasts relatively starkly to the classic economist perspective of dealing with environmental "externalities" that has been developed over the past 30 years. The fundamental argument was that economic measures, including GNP and GDP, did not account for the macro, economic, or social cost of environmental damages. Those damages, where required, were calculated on a private cost basis. Classically, arguments have developed over the decades for "cost of carbon" or "cost of environmental ecosystem services." The calculus behind this, as well as the philosophical approach to these, has been complicated, to say the least.

From a business leadership perspective, there has been an increasingly important recognition of the critical role that businesses and business leaders play in this energy transition, and more broadly in advancing sustainable approaches and solutions. Author and business management consultant John Elkington initiated many of these concepts decades ago – such as the triple bottom line, which has morphed into multiple related threads such as "green business," ESG, and B corps. All of these incorporate the fundamental principles that businesses can and should be prime movers – within the enabling environments that governments create – for both private and social progress. This is what Elkington describes as aligned system value, in which business, society, and environment align together as opposed to being seen in three contrasting systems in which one system imposes losses on another. There are few enlightened business leaders who have embraced these fundamental core principles; if there are business leaders leading a profitable business and solving societal problems while also taking care of the communities and the employees that they work with, and their shareholders, that it is a win-win-win situation. Others, in contrast, may still embrace an increasingly out-of-date paradigm in which businesses should not be responsible for their "externalities," which are instead considered responsibilities of government to regulate and/or mitigate. Thankfully, a new generation of business leaders, in particular those embracing these core philosophies and incorporating fundamentals of value-based businesses that recognize the critical role of employees, working among communities, and working to realize and achieve a regenerative ecosystem, is emerging with increasing scale and impact.[128]

At the intergovernmental level, there have been a few areas of progress to incorporate environmental sustainability into economic analysis, assessments, and policies, and as noted earlier a few attempts to define "green GDP" and infuse that into economic reporting standards. More often than not, informed by the economics community, governments have adopted an approach that values GHG emissions. This was based on fundamental economic paradigms of GHG emissions transitioning from externalities to internalities, that is, to be included in the economic calculus. There has been a large body of literature that argued for economy-wide carbon pricing. This led to the emissions trading system (ETS) in the European Union, the establishment of a carbon trading system in China, and the adaptation of the social cost of carbon into policy analysis in the U.S. There have also been multiple subnational efforts toward establishing a mechanism for incorporating the cost of emissions into full-cost accounting decision-making, such as the regional greenhouse gas initiative in the northeast of the U.S. and carbon pricing in California. These approaches, adopted within a political economy context, strive to introduce the principle but may not in fact have sufficient economic consequences to move markets. For example, the European ETS for years traded at around €10 per ton of carbon dioxide when nearly all of the analysis indicated that carbon pricing above €50 to €100 per ton is needed in order to materially change the economic calculus for alternative technologies to be cost competitive. Only in December 2022 did the European ETS carbon price move north of €80 per ton. This is an additional element of a large policy portfolio including the European Green Deal and many other elements which are steering the European economies toward zero-carbon emissions.

However, it is clear from the academic literature and perhaps more so from the wide body of corporate performance reporting that the economy-wide impacts of pursuing and investing in environmentally benign energy, materials, and processes can have significant positive benefits, including reduced local air pollution, improved ecological productivity, increased biodiversity, and reduced morbidity and mortality, which directly, even according to traditional economics, impact national economic conditions.[129] That is, there needs to be significant continued effort by leading macroeconomists at the highest levels to drive for an equitable and comprehensive economic measurement, metric, and reporting system that includes the interfaces between human systems and earth systems.

The macroeconomic perspective and the systems thinking perspective extend, with some modification, to corporate competitiveness and investors in capital markets. The interconnectivity at play today is through carbon disclosure requirements, reporting requirements, or criteria from investors in the capital markets. There has been complimentary movement from the international accounting standards perspective, as well as the central banks that are the foundation of most banking systems around the world. Here, the UN's Sustainable Finance Initiative has been advancing standards within central banks to account for, at a minimum, carbon emissions and climate change, and in some cases, more comprehensively, broader environmental impacts into their lending and disclosure requirements.[130] Private capital markets are working in parallel, including sovereign wealth funds, private equity and, increasingly, rating agencies and securities and exchange commissions, which promulgate rules for reporting and disclosure.

In the international arena, including the United Nations Framework Convention on Climate Change (UNFCCC), there is a question about the role of government financing to support clean-energy transition as well as climate adaptation. Here, there are two fundamental debates. First, there is a debate among countries regarding impacts and damages. There is a strong voice demanding that those countries which have been emitting GHGs for many decades pay the cost of the clean-energy transition to those countries which have emitted much less GHGs to date and which have much less prosperous economies. The debate stems over the cost-effectiveness of mitigation and adaptation, which some argue cost less than "GHG-intense" and less resilient development pathways. Second, there are ongoing discussions regarding the role of government financing, and in particular the levers which can be turned on in both emerging economies and developed economies. For example, what is the appropriate role of multilateral investment bank financing through other governments as a means of providing concessional finance, grants, local-based capital, or currency, or other risk mitigation mechanisms? These debates contrast with mechanisms being considered in developed economies, such as tax credits, carbon pricing, carbon trade mechanisms, as well as potential or direct cash, grants, and other subsidies for clean-energy transition. The solutions in each country will likely be bespoke, reflecting the circumstances and ambitions of those leaders, as well as their negotiating prowess. They will also reflect the relative investment environment that they create in order to attract the tens of trillions of private capital that is required, and the relatively small amounts of government direct funding that can be deployed. That is, there needs to be a fundamental rethinking and continued advancement of the understanding and the mechanisms for government for creating and enabling risk-adjusted returns to attract private investment, support the necessary innovation, and create fair and transparent rules of engagement for the private sector to thrive. This synergistic ecosystem is critical for implementing the ambitions and goals that have been set out not only by the government but also by the private sector. We must continue to advance our understanding and innovate mechanisms, including sector reforms, contract structures, and financial innovations that can address the challenges of reducing use of (or adding CCS to) fossil fuel infrastructure while simultaneously investing in low-carbon energy solutions through private sector investments that access the local, regional, and global capital markets.

Policy/Regulatory (Local, National, and Global)

The transition into Phase IV, and certainly beyond into a net-zero world, must take into account a different set of dynamics. Where in the past century of the industrial revolution geopolitical considerations were defined by a complex dynamic of governance systems and resource availability, it is evolving dynamically to one that considers much more so the above-ground resources and diminishes the importance of extractive industries, in particular oil, gas, and coal (but perhaps increasingly so for minerals). Resources, however, remain a significant part, particularly energy supply. The role of basic materials – such as silicon, lithium, cobalt, critical minerals that go into magnets, high-power electronics, and batteries, and catalysts – has become increasingly crucial to this energy transition. This will last well into the middle of this century and perhaps beyond depending upon our ability to innovate and create fully recyclable ecosystems. Here, it is important to reemphasize the role of green chemicals and fuels, including hydrogen and ammonia, and how those may play out in the net-zero world. This is consistent with the overall trajectory of energy resource needs and uses.

New dynamics, however, are front of mind, including minerals for the production of renewable energy resources, electronics, batteries, and other clean-energy technologies. The geopolitical dynamics of processing and supply chain flows are increasingly influencing the political policy and investment environment as well as macro thinking at the country level and the corporate level. As many analysts and stakeholders have noted, the components of the U.S. energy policy, which are strongly supportive of domestic mining and manufacturing as well as rapid deployment of clean-energy solutions, are a poignant example of a country's intent to reshape the global supply chains and position for competitiveness in a transitioning clean-energy and net-zero-energy future. In global dialogues, it is increasingly apparent that there are reactionary efforts under consideration across the world including in Canada, Australia, Europe, and throughout Latin America.

One of the more interesting geopolitical dynamics playing out is the debate over power grid interconnectivity. Here, there is political tension due to not only the potential benefits of greater macro grids but also the potential risks of imports and interconnectivity with possibly nefarious actors. These dynamics will continue to play out over the decades – and interestingly open up an innovation landscape for distributed energy solutions. One example of this was the energy transition of New York spearheaded by Richard Kauffman by establishing the New York Green Bank and leading, with others, the New York Reforming the Energy Vision (REV) process. Richard and his colleagues created the framework and implemented "non-wire alternatives" which were set to deploy locational, specific clean-energy system solutions to avoid the need for greater generation and transmission while simultaneously achieving decarbonization, resiliency, reliability, and affordability goals.[131] This was a central part of a large portfolio of actions initiated to decarbonize New York's energy economy.

As noted, while still relatively immature, there is increasing awareness of the need for fully resilient infrastructure planning. Here, "fully" refers to both human and physical system resiliency, including climate resiliency. Interestingly, the IPCC introduced this in AR5 as a concept and it has been more fully fleshed out by the IPCC community in AR6. The concept is well captured in Figure 5 as published recently.[129,132]

However, this more comprehensive framework of resliency remains relatively buried in the academic literature and rarely is front of mind within political and institutional leadership dialogues. Some of this stems from the complexity of working across multiple sectors and also considering energy environmental justice as well as energy access in combination with geopolitics, domestic policy, etc. Solving the complexities at scale may be implemented at the national leadership level with coordination among the various agencies and sectors, allowing them to implement according to their legislative authorities and abilities, or perhaps at the local level, where mayors or governors (or equivalent positions globally) have purview of the full economy of the jurisdiction at what might be considered a "more manageable scale."

At the global level, this complements (and sometimes contrasts with) the desire to achieve the SDGs. This plays out very differently depending on the perspective of the particular person. For example, emerging economy leaders are strongly committed to navigating both economic development and transitioning toward a net-zero energy system, but not at the expense of economic development; thus, many are increasingly negotiating for intergovernmental support and positioning competitively in this transition to a clean-energy economy. Some argue that much like telephony, there is an attractive proposition for leapfrogging the energy infrastructure to one of low carbon that is fully resilient. Some of the arguments for doing so are attractive to local leaders whose economies are organized in sectors which compete for resources and in which they have seen their developed country counterparts be quite successful using high-GHG-emitting approaches that have been considered "low cost." Rarely are political leaders brave enough to focus on longer-term macroeconomic perspectives and incorporate complex intersectoral thinking and perspectives – but, on the positive side, this is emerging, particularly after the outbreak of the Ukraine war and as codified within the

From climate risk to climate resilient development: climate, ecosystems (including biodiversity) and human society as coupled systems

(a) Main interactions and trends

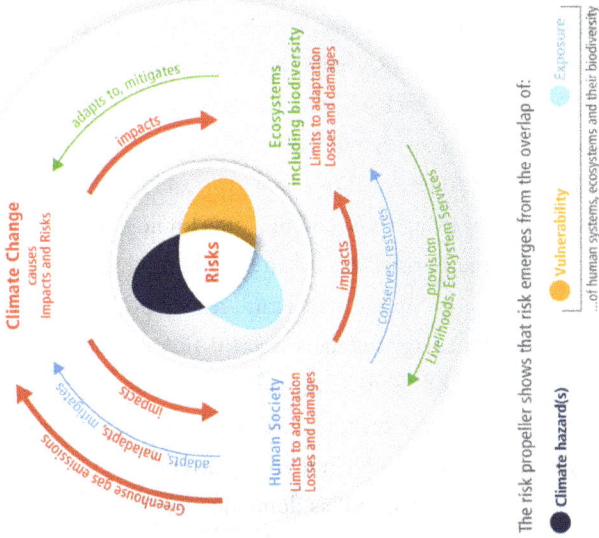

(b) Options to reduce climate risks and establish resilience

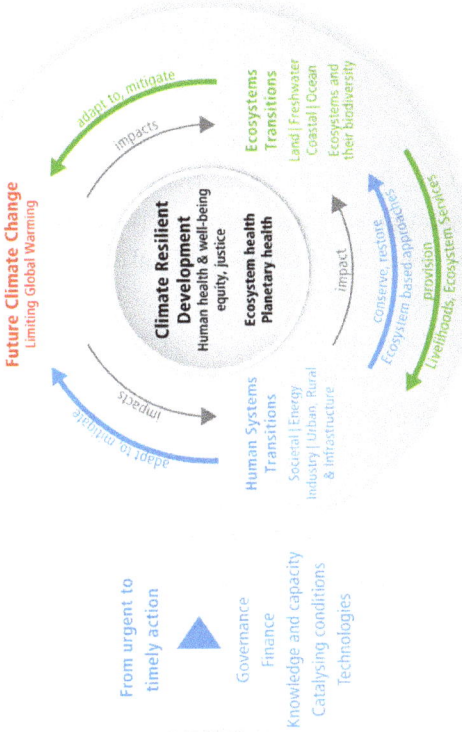

The risk propeller shows that risk emerges from the overlap of:

Climate hazard(s) | Vulnerability | Exposure

...of human systems, ecosystems and their biodiversity

Figure 5. *Transition from risk to climate resilient development framework.*
Source: Reprinted with permission from IPCC.

policy packages of the U.S. and Europe. The opportunity is for the economists, geopolitical thought leaders, complex system thinkers, and leaders to work with governments, non-profits, and industry to advance the value-added complexities across the globe.

At the national level, multiple countries have initiated and put in place very strong policy environments to support the energy transition, notably the U.S. and countries of the European Union (among many others such as the U.K., Australia, and New Zealand). The European Union's Green Deal is the Union's commitment to achieving net-zero emissions by 2050. It encompasses not only the elements for the energy transition but also provisions to attempt to decouple economic growth from resource use, including circular economy strategies and energy and environmental equity provisions. It also includes a significant amount of investment and strong sector-specific targets. For example, in transportation, the goal is to reduce emissions by 55% from cars by 2030, by 50% from vans by 2030, and have zero emissions from new cars by 2035. That also includes a target for 40% renewable energy by 2030, and 36–39% of new energy efficiency targets for final and primary energy consumption, which would be predominantly realized in building energy efficiency and industrial efficiency.[133]

At the national level, the U.S. passed what is considered to be landmark legislation in 2022; the most significant piece of this legislative package was the Inflation Reduction Act (IRA). The IRA built upon a series of energy-related policies, with recent major acts going back to 1992, complemented by the energy policy act of 2005, the American Recovery and Reinvestment Act of 2009, and the Energy Act of 2020. In 2022, the Infrastructure, Investment, and Jobs Act (IIJA) laid important groundwork for investing in clean-energy technologies as part of COVID-19 recovery. The IRA authorized more than $1 trillion in new spending with nearly $550 billion for energy- and infrastructure-related projects. Further, it included nearly $75 billion for energy and materials. The IRA contains numerous provisions for the advancement of clean-energy solutions stemming from carbon capture utilization, continuation of the PTC and ITC for renewables, and storage of low-carbon hydrogen. Provisions include strong incentives to reduce the lifecycle of greenhouse gas emissions as well as domestic content. Nearly all the provisions are considered "carrots" versus "sticks"; that is, they provide tax incentives, grants, or other fiscal incentives for development and production of the technologies and solutions that are sought to support the transition to a low-carbon economy versus regulatory requirements or penalties for not attaining goals. They also complement elements which were included in the Creating Helpful Incentives to Produce Semiconductors (CHIPS) Act, which appropriated $52.7 billion, and has multiple elements to build domestic supply chains.[134] Additional elements are in the IRA, including incentives for the development of new critical minerals, lithium mining, battery manufacturing, PV manufacturing, electric vehicle manufacturing, and other elements of the green economy. These national legislation packages are over and above additional appropriations, which flow through the various U.S. government agencies, including the U.S. Department of Energy and the Treasury Department, to implement the research, development, demonstration and deployment (RDD&D) programs, as well as the technical assistance and support to states, low-income communities, and other jurisdictions.

Many around the world do not see the IRA in such a positive light. Many see it as an industrial policy wrapped into a climate and decarbonization agenda. Others see it as an "anti-China" policy for clean-energy technologies. There is ongoing discussion on the development of a policy response from the European Union that has initially been codified in the European Green Deal. Others see it as an initial step in the resetting of the globalization agenda and positioning for the geopolitics of a low-carbon energy economy.[135] This has considerable implications for the next several decades. Ongoing discussions include the creation of alliances for the development of geopolitically resilient supply chains and coalitions to build economies of scale to rapidly reduce the cost of critical technologies for a net-zero economy, including electrolyzers, fuel cells, batteries, expansion of solar power, wind power, and carbon capture and storage, as well as conversion technologies.[136]

While climate and related energy programs garner much attention in the U.S. and Europe, other nations have also been putting provisions in place, such as India and China. Both those countries have laid out roadmaps for longer-term transitions to a low-carbon economy, with 2060–2075 time frames, and have also laid out aggressive targets for renewable electricity generation and growth, building of domestic supply chains and more resilient supply chain alliances, as well as transitioning mobility and investing in energy efficiency.[137,138] Other countries such as Canada, Chile, and Australia have also come forward with strong national targets and policy packages toward a green economy, including strong provisions for competitiveness for green hydrogen and green chemicals.[139-141] One can see the evolving global dynamics for competitiveness begin to take shape within the green economy. This will play out very dynamically over the next two to three decades.

The portfolio of activities that exist at the national level is sometimes complimented by activities at the state or local levels. In the European Union (EU), the EU targets translate to member country targets, as well as additional policy packages. In the U.S., there are additional state policies as well as local policy packages that are driving transformation. Some of these were described in the earlier Phases. They continue to evolve and expand. Today and going forward, they are strongly complimented by national-level activities, including the RDD&D, grants, loan programs, and provisions of the IRA bill.[c]

Knowledge (Models, Tools & Data for Planning & Operations and Business Models)

In Phase IV, knowledge has three principal focus areas. First, a few persistent myths and paradigms with knowledge that were learned back in Phases II and III must be addressed. Many older-generation leaders retain these out-of-date perspectives in their current roles and represent significant barriers to innovation and adaptation of sustainable technologies and solutions.

Second, new knowledge must be advanced, incorporating effective communication strategies, to address these inherent human, social, and institutional barriers. These are all in addition to the technology innovations which were outlined earlier, including expansion of knowledge which is under development for grid-forming inverters, operational aspects and technology needs for high-penetration variable renewable systems, control and operations of millions of interconnected devices, such as DERs, batteries, EVs, and HEMS, and hierarchical distributed controls

---

c   Up-to-date information on state policies in the U.S. is available at dsireusa.org.

enabled by AI/ML and cybersecure cloud computing. Third, it is increasingly important that the social license to operate, addressing energy and environmental justice concerns, and advancing additional social and institutional knowledge, be effectively addressed to realize the solutions at speed and scale across the globe. This implies that the incorporation and advancement of a broader range of social sciences will be increasingly important. Over the past few years, there has been increased awareness of, and now emphasis on, just transitions. This includes multiple aspects of procedural and inclusive processes. It is also increasingly recognized, with the availability of very large datasets and the use of AI, that there is a growing opportunity to build new knowledge and approaches toward effecting change at speed and scale. Here, consider building up not only scientific, business, and economic innovations and knowledge but also the social sciences aspects of inclusive processes, use of social media, communications methodologies, etc. All of these factors must come into play during Phase IV in order to effect change at the speed and scale that is necessary.

It is important to separate economics from other disciplines of social sciences. As discussed previously, the evolution in economic modeling and analysis will need to develop to take into account social and environmental justice. This evolution in capabilities must also incorporate innovative methods and approaches to include the costs and benefits of environmental attributes of solutions, alternative approaches to geopolitical security, resiliency, and more. All of that falls within the umbrella of advancing knowledge as part of our transition to a net-zero economy.

Finally, we laid out in the above section, from a governance policy and regulatory leadership perspective, the critical need to advance our knowledge, supported by tools, data, and further innovation in creative thinking. This will require expanding our ability to measure, translate into micro- and macroeconomic tools, and critically evaluate the interconnectivity and systemic transformations that need to take place over the next several decades. We need to build not only the conceptual framework but also implement practical examples that can be touched, felt, and experienced by not only those who live there but also those who visit. These examples for realization are critical to breaking open the bounds of thinking that constrain creative problem-solving. Additionally, we need to advance the critical interfaces among academics and practitioners from different sectors. For example, we need much tighter integration between those who study climate and those who study the energy and natural resources systems. We need deeper integration between social scientists and physical scientists and much tighter integration between scientists, engineers, and those who excel in communicating complexities to others. We need to communicate, likely through visualizations that the brain processes much more easily than reports or graphs, to convey complex systems in ways that help build cognitive comprehension (e.g., visual learning) to deepen the appreciation and opportunities for multisectoral, multipurpose solutions. Combining storytelling with gaming and meta-verse immersion environments would be beneficial in creating greater intuition to break down the barriers among sectors and offer creative problem-solving and systemic multisectoral solutions.

One implementation approach has been suggested as a means to build effective intuition for the values of renewable-based energy economies, at least for those techno-economically oriented: One may reimagine the next generation of "Sim City" in which we can create, explore, and evaluate resilient, net-zero solutions for communities, cities, regions, states, or nations. Bringing those technically rigorous intuition-enabling tools to fruition could offer enormous advances for this transitional pathway. Of course, that needs to be combined with sophisticated understanding and approaches to the economics, the politics, and the political–economic dynamics of moving in those directions.

Social Willingness

There are many aspects to this broad category of "social willingness," including popular sentiment, the influence of non-governmental organizations and advocacy organizations, the role of information and disinformation, the role of government in informing and leading, social change, and the role of community. In the overall impact factor diagram (Fig. 1) and the figure model that we have been following throughout these Four Phases, social willingness is quite strong in Phase IV, but not overwhelming on average. In some jurisdictions such as the European Union, social willingness is very strong. In other locations, such as the U.S., Africa, Southeast Asia, Russia, and China, it is more difficult to assess the overall sentiment, and the distribution of sentiment is much broader than in the European Union. However, on a generational basis, it is quite interesting to note that generations from the millennials through Gen X, Y, and Z, who are much more social media aware and connected, as well as much more aware of environmental degradation and its impact on their lives, have a much narrower distribution and are more positively inclined to support an overall clean-energy transition and a transition to a net-zero energy economy.[142] One can see this play out not only in the media but also across schools and university campuses, as well as throughout small and creative, socially minded business environments. All this bodes positively for future generations to continue to assert strong leadership throughout the various channels in which they engage socially, in business and their community, and in government.

Additionally, there is a renewed concentration of activity to address the SDGs, in particular, SDG 7, which focuses on energy access. New efforts are being made to take fresh approaches to this, for example, the Global Energy Alliance for People and Planet (GEAPP), which is a partnership of the Rockefeller Foundation, the Bezos Foundation, and the Ikea Foundation.[143] The GEAPP not only has commitments of more than $1 billion from the foundations but has also created a strategy to leverage that into hundreds of billions of dollars to lead transformational projects to address energy access as well as the energy transition. It is efforts like this, in combination with a plethora of efforts led out of coalitions such as the Energy Transitions Commission,[144] the World Economic Forum,[145] and Clean Energy Ministerial,[146] that bode well for building continued momentum and social willingness to support this energy transition.

The increase in social willingness offers a unique opportunity for collaboration with leading academics, policy innovators, and economists to take a fresh approach toward "quality-of-life" solutions versus singular sectoral solutions. Here, for example, one might reflect on the work described earlier on human-centered transitions. Expanding on that, particularly in the post-COVID-19 environment, one can readily reflect on the quality-of-life contributions that come from comfort, connectivity, mobility, and the provision and consumption of goods. Think of a roof over one's head with thermal comfort, that is, heating and cooling to be physically comfortable. Connectivity, of course, has changed dramatically since the COVID-19 pandemic with considerably more Internet and video connectivity replacing travel and in-person work or meetings. The ubiquity of smart phones and Internet access, at affordable prices, offers a potentially transformational opportunity to continue to provide a high quality of life at lower energy consumption as well as lower GHG footprints for everyone. Some may wish to extend this all the way into concepts of the meta-verse. Perspectives on mobility and goods have also evolved, transitioning to considerations of the GHG footprint of the mode of travel – that is, electric vehicle versus train versus airplane versus walking – and goods coming from either wasteful and environmentally impactful supply chains and processes or more local, environmentally benign, organic, and low-GHG-footprint supply chains and processes. This extends into an increasing consciousness of the GHG footprint of diet.[147] On this, there is an increasing awareness of the heavy GHG footprint of the approach of the last century, which was to cut forests for pastureland, to graze cattle or other animals and process them to deliver high-value proteins. Similarly, intense fishing of the oceans has led to increased consciousness of the intense depletion of fish populations and the health of the oceans, which need to be replenished and rejuvenated. On a technological front, multiple companies have sprung forward to deliver plant-based proteins as alternatives, and there is a growing movement toward vegetarian and vegan diets. Whether or not this plays out in all regions of the world is yet to be seen. And, the overall effectiveness of these movements is also not well known.

Institutional Willingness and Political Economy Dynamics

Institutional willingness reflects the overall dynamics of corporate, government, and international institutions, and their reflective interest and ability to drive forward additional ambitions for as well as implementation of clean-energy transition. In the past, many of these dynamics have played out on individual stages. One example is the international negotiation that took place under the UNFCCC among countries, particularly during the Copenhagen to Paris Conference of Parties (COPs) transition, and the realization of the Paris Agreement, which brought forward the recognition that non-binding but politically influential sets of commitments could be a potentially effective pathway. Within that approach, countries have both different levels of ambition and different time frames, even though they have agreed to a cadence for assessing progress and reevaluating their

NDCs, or the depths of their commitment and the time frames for that. Under U.S. Special Envoy John Kerry in the last few years, we have seen increased political drive for both implementation and raising ambition. This has come with a tremendous amount of diplomacy, combined with sophisticated approaches and programs for public–private partnerships, financing, technical assistance, and other elements to support the transition.

We previously reflected on the orientation of many nation states as well as the growing movement among major corporations to drive or support the clean-energy transition and net-zero economy transition. At the institutional level, regulatory and permitting processes are of critical concern for the speed and ability to develop new, clean-energy projects. For example, there is growing concern that there will be bottlenecks and time constraints to permit new transmission lines in the U.S., or potentially in Europe. Likewise, identifying and permitting project development sites, particularly with access to existing or new transmission lines, represent potentially major concerns. New net-zero industrial clusters also need permits for the development of those sites. Pipelines and conversion plants for hydrogen to ammonia, for example, or for shipping and processing of ammonia or other green chemicals, may also need new infrastructure and new permits to build. This institutional willingness – as well as the more local regulatory bodies, which, in the U.S. as an example, hold considerable sway over state utilities – is critical and in need of capacity building, knowledge, and awareness. Further, they would greatly benefit from new processes, data, availability, and IT provisions (such as natural-language-assisted processing of permitting applications). That is, we may need considerable new innovation to bring IT, AI, cloud computing, and electronic processes and efficiencies to the realization pipeline of institutional processes in order to effectively build a new energy economy.

From a political economy standpoint, we also see potential risks. Elected officials, corporate leaders, non-governmental leaders, and other influencers are affected by the power that they wield. In this realm, we have to be aware of, strategize pathways to address, and implement through creative coalitions the political economy dynamics. Navigating this will require not only critical insights but also diplomatic savvy, impactful storytelling, and network building that perhaps have not been attempted successfully in the past.

As we reflect on the critical role that the private sector plays in bringing capital to bear, in technology and business model innovation, in employment, and in influence on future generations, the political economy, and government, it is also valuable to reflect on the approaches toward understanding business leadership. A number of years ago, a colleague and I conducted multiple C-suite interviews, reflecting on the new paradigm for business innovation in the transition to a green energy economy. We leveraged the classic approach of Michael Porter's Five Forces, and updated that given the new set of dynamics, as well as the evolving geopolitical concerns and the increased awareness of the need to account for climate and other extreme events and threats.[148] The updated approach, shown in Figure 6, incorporates five primary drivers or forces for business, innovation, and strategy:

(1) **Data-driven and IT-enabled solutions:** This incorporates the elements that were described earlier, including massive data availability, AI/ML, cloud computing, ubiquitous available Internet, and automated processes.

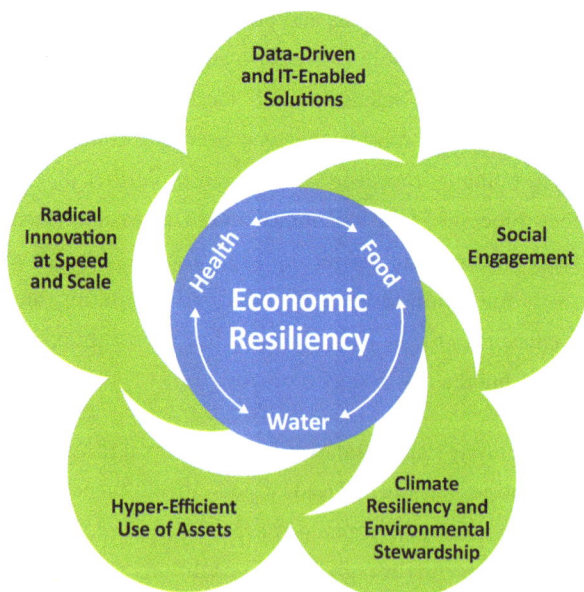

Figure 6.   21st Century Innovation Strategy. New strategic areas that leaders must understand and navigate to be successful as the world transitions to sustainable energy systems and overall net-positive dynamics.

(2) **Social engagement:** This includes factors that corporate leaders must consider regarding engagement with their evolving workforce, their communities, their investors, the larger community of non-governmental organizations, and governments. No longer are corporations solely responsible for delivering financial returns to their investors. Smart leaders recognize that they must embrace a well-thought-through social engagement strategy in order to effectively lead a corporate enterprise which depends on the license to operate as well as culture and employee well-being. Some companies have embraced this in a new senior executive role, which is beyond public relations or human resources, and consider the corporation's role in what is called "human flourishing."

(3) **Physical and cyber resiliency, and environmental stewardship:** This factor incorporates the strategies that must be considered for investment, R&D, near-term operations, as well as long-term resiliency, particularly within an overall transition to a net-zero economy and circularity. Companies that do not embrace these elements face physical or operational risk, such as losing their license to operate or facing severe penalties for lack of environmental stewardship. All of these translate directly to both top-line and bottom-line impacts and must be considered as part of the strategic framework.

(4) **Hyper-efficient use of assets:** Here, there is a fundamental rethinking of asset investments, balance sheets, and utilization of others' capital in creating service business models and platforms. Further, for infrastructure projects, flexible utilization of assets in poly-generational – that is, risk-adjusted inputs and multiple revenue streams – integrated facilities will become increasingly differentiating. Mapping out the strategic alternatives and hedging among options, as appropriate for each operating jurisdiction, will be critical to maximizing the return on investment and developing highly scalable, flexible business operations.

(5) **Radical innovation at speed and scale:** It used to be that businesses could define a core value proposition and a core set of assets, products, and services to sell to customers. Innovation refined and continued to evolve those product lines. Threats to innovation came from new entrants into the sector. That paradigm has changed over the last few decades, in which companies have had to embrace the strategy of innovation at speed and scale, potentially remaking themselves, as well as looking much more broadly into

the innovation ecosystem to seek new acquisitions and ideas to scale through strategic investment. This should also include embracing new approaches to feeding the pipeline for innovation, such as engagements with universities and other start-up incubators throughout the world, approaching applied engineering facilities, such as the national laboratories, and devising a strategy for leveraging those investments in a highly effective and efficient manner. All this must be structured within a strategy which recognizes that the new innovations are coming at a much faster pace, and must be brought to market at a speed and scale that are unprecedented.

The operating context for these five strategic thrusts continues to evolve. In Figure 6, we have outlined the interconnectivity between health, food, water, and economic resiliency. These elements reflect the interconnectivity in the evolving geopolitical landscape, while also reflecting the roles that businesses play in the provision of critical products and services for health, clean water, food, and economic resiliency through the provision of power, energy, employee wellness programs, preparedness for dealing with disasters, and more. Business leaders such as those who are embracing green swan strategies, as John Elkington refers to them, are crucial to creating the public–private partnerships, the private–private partnerships, and the innovation ecosystems, and bringing to bear the investment that will be needed in order to realize this clean-energy and sustainable future.

Chapter 5
Phase V: *Beyond 2050 –*
A Global Net-Zero Energy
Economy (2050+)

It is fascinating to reflect on the past Phases of our energy transition, in particular the enormous growth in technical capabilities, knowledge, social and political willingness, and economic competitiveness, which have led to Phase IV, which we are in now. This transformation and evolution toward a future with renewable energy and a net-zero economy holds enormous promise to achieve the bold science-based targets and ambitions that have been laid out by governments, corporations, and societies around the world. It will not be a straightforward or easy task. However, I remain cautiously optimistic.

As noted throughout Phase IV, the needed investment for the energy transition is on the order of one hundred trillion dollars over the coming decades and entails building out 3x–10x our clean power system and transforming nearly all of our energy infrastructure to become sustainable (low/zero GHG emissions, low water use, and low/zero environmental impact). Global ambitions for advancing renewable energy and energy efficiency were formalized in December 2023 at the Conference of Parties (COP) in which more than 120 countries signed the Global Renewable Energy and Energy Efficiency Pledge. The Pledge succinctly declared the government's intent to work collaboratively, as rapidly as possible, to triple renewable energy generation to at least 11 TW by 2030 and double energy efficiency.[149] This will not be straightforward. There are challenges related to land use, social acceptance, material availability and related geopolitical above-ground issues, changing politics, and economics. From a social perspective, there is an enormous amount of work required for positive transitions for not only those

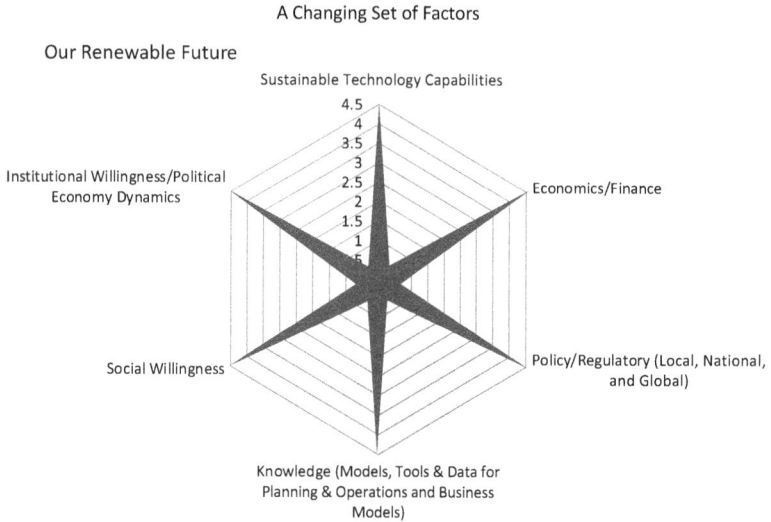

Figure 1. A changing set of factors for Phase V.

without access to modern energy but also for those who have been previously negatively affected by the energy infrastructure over the past many decades. This includes those impacted by environmental and health aspects and those who will require positive job transitions. Implementing this at the scales required in the U.S., throughout Europe, and across the world will take concerted efforts of leaders across governments, non-profits, and industry. While an enormous challenge, I remain hopeful that there are numerous leaders who have led these transitions in the past decade, and their lessons and stories must be disseminated around the world and translated into the local, social, and political contexts.

Looking beyond Phase IV of this renewable-energy-dominated future to Phase V, in which the world is operating as a global net-zero economy, we see an enormous landscape for positive innovation, technological change, social change, geopolitical dynamics, and achievement of the SDGs without compromise, as depicted in Figure 1. What will likely still be at play will be continued positioning for political power and potentially religious-based strife and wars, which hopefully will reduce within this much more enlightened future. A number of ideas that are still under development are likely to mature over the coming decades and have an impact on

a much more substantial scale, including global carbon-based pricing of goods and services; global trade of clean chemicals and fuels; a new global reuse, recycle, and repurpose circular economy; and new businesses which understand that their roles are beyond a singular sector. Here, we may see an increase of conglomerates, for example, of strategic alliances across the energy, food, water, and infrastructure sectors. These strategic alliances or conglomerates would be positioned to take advantage of the strategic synergies among and across sectors, as opposed to seeing them as just handoffs of goods, services, and money between the sectors.

Final Reflections and Our Call to Action

As we are only entering Phase IV now, and given the current GHG footprint and energy mix of the world, there is considerable work to be done. While I have laid out the potential for realization of this new energy economy, it is up to us as students, teachers, practitioners, and leaders in all sectors to embrace the challenge in front of us and work together to break down the barriers and implement change at the speed and scale required for the energy and overall economic transition – for us, for our children, for our grandchildren, and for the next seven generations.

Appendix A
List of Publications by the Author (as of June 2023)

Adams, R., Pless, J., Arent, D. J., *et al.* (2016). *Accelerating Clean Energy Commercialization. A Strategic Partnership Approach.* Golden, CO, USA: National Renewable Energy Laboratory (NREL). https://doi.org/10.2172/1249272.

Ahamed, S., Sperling, J., Galford, G., *et al.* (2019). The food-energy-water nexus, regional sustainability, and hydraulic fracturing: An integrated assessment of the Denver Region. *Case Studies in the Environment,* 3(1), 1–21. https://doi.org/10.1525/cse.2018.001735.

Arent, D. J. (1987). Interfacial Charge Transfer Properties of the N-Cadmium .X.; .X = Sulfur, Selenium, Tellurium. Semiconductor in Contact with Metal .X. Iron Hexacyanide .4-; 3-.; .Metal= Lithium, Sodium, Potassium, Rubidium, Cesium. Electrolytes and AC Impedance Studies of the Nickel Hexacyanoferrate .2-, 1-. Modified Electrode. Princeton University.

Arent, D. J. (2010). Energy: A national and global issue. In H. Cabayan, B. Sotirin, R. Davis, *et al.* (eds.), *Perspectives on Political and Social Regional Stability Impacted by Global Crises — A Social Science Context.* Strategic Multilayer Assessment (SMA) and U.S. Army Corps of Engineers Research and Development Directorate. Available at https://nsiteam.com/social/wp-content/uploads/2016/01/Perspectives-on-Political-and-Social-Regional-Stability-Impacted-by-Global-Crises-A-Social-Science-Context.pdf.

Arent, D. J. (2010). The role of renewable energy technologies in limiting climate change. *Bridge,* 40(3), 31–39. Available at https://www.nae.edu/24531/The-Role-of-Renewable-Energy-Technologies-in-Limiting-Climate-Change.

Arent, D. J. (2016). After Paris, the smart bet is on a clean energy future. *Greenmoney,* USA.

Arent, D. J. and Bocarsly, A. B. (1987). The influence of aqueous alkali cations on the electrochemistry of cadmium ferricyanide overlayers on N-Cdx X = S, Se, Te. Photoanodes. *Abstracts of Papers of the American Chemical Society*, p. 191.

Arent, D. J. and Sweeney, J. (2010). Solar issues. *MIT Technology Review*, p. 113.

Arent, D. J., Arndt, C., Miller, M., *et al.* (eds.). (2017). Introduction and synthesis. In *The Political Economy of Clean Energy Transitions*. Oxford University Press, UK.

Arent, D. J., Arndt, C., Tarp, F., *et al.* (eds.). (2017). *Moving Forward: The Political Economy of Clean Energy Transitions*. Oxford University Press, UK.

Arent, D. J., Amos, L. J., and Bocarsly, A. B. (1992). Structurally dependent properties of the cadmium ferricyanide overlayer in N-Cds and N-Cdse/Fe.Cn.64-/3-photoelectrochemical cells. *Journal of the Electrochemical Society*, 134(3), C144–C144.

Arent, D. J., Balash, P., Boardman, R., *et al.* (July 24–25 2018). *Summary Report of the Tri-Lab Workshop on R&D Pathways for Future Energy Systems*. Golden, CO, USA: National Renewable Energy Labarotory (NREL), NREL/TP-6A20-72926.

Arent, D. J., Barnett, J., Mosey, G., *et al.* (2009). The potential of renewable energy to reduce the dependence of the state of Hawaii on oil. In *2009 42nd Hawaii International Conference on System Sciences*, Waikoloa, HI, USA, pp. 1–11. https://ieeexplore.ieee.org/document/4755531.

Arent, D. J., Barrows, C., Davis, S., *et al.* (2021). Integration of energy systems. *MRS Bulletin*, 46, 1139–1152. https://doi.org/10.1557/s43577-021-00244-8.

Arent, D. J., Benioff, R., Mosey, G., *et al.* (2006). *Energy Sector Market Analysis*. Golden, CO, USA: National Renewable Energy Laboratory (NREL). Available at https://www.nrel.gov/docs/fy07osti/40541.pdf.

Arent, D. J., Bragg-Sitton, S. M., Miller, D. C., *et al.* (2021). Multi-input, multi-output hybrid energy systems. *Joule*, 5(1), 47–58. https://doi.org/10.1016/j.joule.2020.11.004.

Arent, D. J., Denholm, P., Drury, E., *et al.* (2011). Chapter 13 — Prospects for renewable energy. In P. Fereidoon and B.-H. Sioshansi (eds.), *Energy, Sustainability and the Environment* (pp. 367–416). https://doi.org/10.1016/B978-0-12-385136-9.10013-0.

Arent, D. J., Döll, P., Strzepek, K. M., *et al.* (2014). Cross-chapter box on the water–energy–food/feed/fiber nexus as linked to climate change. In *Climate Change 2014: Impacts, Adaptation, and Vulnerability. Part A: Global and Sectoral Aspects. Contribution of Working Group II to the Fifth Assessment Report of the Intergovernmental Panel on Climate Change.* Cambridge University Press, Cambridge, UK and New York, NY, USA, pp. 163–166.

Arent, D. J., Engel-Cox, J., Logan, J., *et al.* (2020). *Next Generation Renewables and, Separately, Energy System Transformation, and Related Topics: Cooperative Research and Development Final Report*, CRADA Number CRD-14-557. National Renewable Energy Laboratory (NREL), Golden, CO, USA, NREL/TP-6A50-76744. https://doi.org/10.2172/1659919.

Arent, D. J., Hidalgo-Luangdilok, C., Chun, J. K. M., *et al.* (1992). The roles of cesium ions in regulating photoinduced interfacial charge transfer at the N-type cadmium chalcogenide/ferrocyanide interface. *Journal of Electroanalytical Chemistry*, 328(1–2), 295–310. https://www.sciencedirect.com/science/article/abs/pii/0022072892801868?via%3Dihub.

Arent, D. J., Logan, J., Macknick, J., *et al.* (2015). A review of water and greenhouse gas impacts of unconventional natural gas development in the United States. *MRS Energy & Sustainability*, 2. https://link.springer.com/article/10.1557/mre.2015.5.

Arent, D. J., Peterson, M. W., Kramer, C., *et al.* (October 1996). Correlation of photoluminescence linewidths with carrier concentration in P-Ga0.52In0.48P. *Journal of Electronic Materials*, (10), 1633–36. https://doi.org/10.1007/bf02655588.

Arent, D. J., Pless, J., and Statwick, P. (2016). *Five Forces of 21st Century Innovation Strategy: Insights for Leaders.* Golden, CO, USA: National Renewable Energy Laboratory (NREL). https://www.osti.gov/servlets/purl/1244311.

Arent, D. J., Pless, J., Mai, T., *et al.* (2014). Implications of high renewable electricity penetration in the US for water use, greenhouse gas emissions, land-use, and materials supply. *Applied Energy*, 123, 368–377. https://doi.org/10.1016/j.apenergy.2013.12.022.

Arent, D. J., Rubin, H. D., Chen, Y., *et al.* (1992). Cadmium ferrocyanide overlayers: Regulation of photoinduced charge transfer at the N-Cdse/[Fe.CN.6] 4-/3-Interface. *Journal of Electrochemical Society*, 139, 2705. https://doi.org/10.1149/1.2068968.

Arent, D. J., Tol, R. S. J., Faust, E., *et al.* (2014). Key economic sectors and services. In *Climate Change 2014: Impacts, Adaptation, and Vulnerability. Part A: Global and Sectoral Aspects. Contribution of Working Group II to the Fifth Assessment Report of the Intergovernmental Panel on Climate Change.* Cambridge University Press, Cambridge, UK and New York, NY, USA, pp. 659–708.

Arent, D. J., Wise, A., and Gelman, R. (July 2011). The status and prospects of renewable energy for combating global warming. *Energy Economics*, (4), 584–93. https://doi.org/10.1016/j.eneco.2010.11.003.

Arndt, C., Arent, D. J., Hartley, F., *et al.* (2019). Faster than you think: Renewable energy and developing countries. *Annual Review of Resource Economics*, 11, 149–168. https://www.annualreviews.org/doi/abs/10.1146/annurev-resource-100518-093759.

Arndt, C. and Arent, D. J. (2016). Special section of applied energy: Energy and climate change in sub-Saharan Africa. *Applied Energy*, 161, 553–555. https://doi.org/10.1016/j.apenergy.2015.10.160.

Badger, J., Jørgensen, H. E., Kelly, M. C., *et al.* (2012). Discovering the true wind resource: Including hi-res terrain effects for a new and global wind atlas. In EWEA 2012-European Wind Energy Conference & Exhibition. Available at https://backend.orbit.dtu.dk/ws/portalfiles/portal/7946873/Discovering_the_true_wind.pdf.

Bazilian, M., Brandt, A. R., Billman, L., *et al.* (2014). Ensuring benefits from North American Shale gas development: Towards a research agenda. *Journal of Unconventional Oil and Gas Resources*, 7, 71–74. https://doi.org/10.1016/j.juogr.2014.01.003.

Bazilian, M., Davis, R., Pienkos, P. T., *et al.* (2013). The energy-water-food nexus through the lens of algal systems. *Industrial Biotechnology*, 9(4), 158–162. https://doi.org/10.1089/ind.2013.1579.

Bazilian, M., Mai, T., Baldwin, S., *et al.* (2014). Decision-making for high renewable electricity futures in the United States. *Energy Strategy Reviews*, 2(3–4), 326–328. https://doi.org/10.1016/j.esr.2013.11.001.

Bazilian, M., Miller, M., Detchon, R., *et al.* (2013). Accelerating the global transformation to 21st century power systems. *The Electricity Journal*, 26(6), 39–51. https://doi.org/10.1016/j.tej.2013.06.005.

Bazilian, M., Onyeji, I., Liebreich, M., *et al.* (2013). Re-considering the economics of photovoltaic power. *Renewable Energy,* 53, 329–338. https://doi.org/10.1016/j.renene.2012.11.029.

Bazilian, M., Rogner, H., Howells, M., *et al.* (December 2011). Considering the energy, water and food nexus: Towards an integrated modelling approach. *Energy Policy,* (12), 7896–7906. https://doi.org/10.1016/j.enpol.2011.09.039.

Blok, K., Hare, W., Hohne, N., *et al.* (2011). How to bridge the gap — What the scenarios and studies say. In *Bridging the Emissions Gap: A UNEP Synthesis Report.* https://wedocs.unep.org/20.500.11822/7996.

Bocarsly, A. B., Arent, D. J., and Amos, L. J. (1998). Interfacial structure as a controlling element in photoinduced charge-transfer in N-CdS and N-Cdse/Fecn64-/3-Cells. *Abstracts of Papers of the American Chemical Society,* 193.

Brandt, A., Heath, G. A., Kort, E. A., *et al.* (2014). Methane leaks from North American natural gas systems. *Science,* 343, 733–735. doi: 10.1126/science.1247045.

Brown, E., Cory, K., and Arent, D. J. (2007). *Understanding and Informing the Policy Environment: State-Level Renewable Fuels Standards.* Golden, CO, USA: National Renewable Energy Laboratory (NREL). Available at https://www.nrel.gov/docs/fy07osti/41075.pdf.

Bush, B., Hanes, R., Hunter, C., *et al.* (2020). *Workshop Report on Methods For R&D Portfolio Analysis and Evaluation.* Golden, CO, USA: National Renewable Energy Laboratory (NREL). https://doi.org/10.2172/1665806.

Bush, B., Jenkin, T., Lipowicz, D., *et al.* (2012). *Variance Analysis of Wind and Natural Gas Generation under Different Market Structures: Some Observations.* Golden, CO: NREL. NREL/TP6A20-52790. Available at http://www.nrel.gov/docs/fy12osti/53730.pdf.

Carmichael, S., Booten, C., Robertson, J., *et al.* (2016). *Annual Energy Savings and Thermal Comfort of Autonomously Heated and Cooled Office Chairs.* Golden, CO, USA: National Renewable Energy Laboratory (NREL). https://www.nrel.gov/docs/fy16osti/66431.pdf.

Cochran, J., Bird, L., Heeter, J., *et al.* (2012). *Integrating Variable Renewable Energy in Electric Power Markets. Best Practices From International Experience, Summary For Policymakers.* Golden, CO, USA: National Renewable Energy Laboratory (NREL). Available at https://www.nrel.gov/docs/fy12osti/53730.pdf.

Cochran, J., Bird, L., Heeter, J., *et al.* (2012). *Integrating Variable Renewable Energy in Electric Power Markets: Best Practices from International Experience.* Golden, CO, USA: National Renewable Energy Laboratory (NREL). https://doi.org/10.2172/1041369.

Cochran, J., Miller, M., Milligan, M., *et al.* (2013). *Market Evolution: Wholesale Electricity Market Design for 21st Century Power Systems.* Golden, CO, USA: National Renewable Energy Laboratory (NREL). Available at https://www.nrel.gov/docs/fy14osti/57477.pdf.

Cochran, J., Miller, M., Zinaman, O., *et al.* (2014). *Flexibility in 21st Century Power Systems.* Golden, CO, USA: National Renewable Energy Laboratory (NREL). https://doi.org/10.2172/1130630.

Cochran, J., Zinaman, O., Logan, J., *et al.* (2014). *Exploring the Potential Business Case for Synergies between Natural Gas and Renewable Energy.* Golden, CO, USA: National Renewable Energy Laboratory (NREL), NREL/TP-6A50-60052. Available at https://www.nrel.gov/docs/fy14osti/60052.pdf.

Davis, S. J., Lewis, N. S., Shaner, M., *et al.* (2018). Net-zero emissions energy systems. *Science, 360*(6396), 9793. https://doi.org/10.1126/science.aas9793.

Denholm, P., Arent, D. J., Baldwin, S. F., *et al.* (2021). The challenges of achieving a 100% renewable electricity system in the United States. *Joule, 5*(6), 1331–1352. https://doi.org/10.1016/j.joule.2021.03.028.

Elishav, O., Mosevitzky Lis, B., Miller, E. M., *et al.* (2020). Progress and prospective of nitrogen-based alternative fuels. *Chemical Reviews, 120*(12), 5352–5436. https://doi.org/10.1021/acs.chemrev.9b00538.

Ericson, S., Anderson, K., Engel-Cox, J., *et al.* (2018). Power couples: The synergy value of battery-generator hybrids. *The Electricity Journal, 31*(1), 51–56. https://doi.org/10.1016/j.tej.2017.12.003.

Ericson, S. J., Engel-Cox, J., and Arent, D. J. (2019). *Approaches for Integrating Renewable Energy Technologies in Oil And Gas Operations.* Golden, CO, USA: National Renewable Energy Laboratory (NREL), NREL/TP-6A50-72842.

Ericson, S. J., Rose, E., Jayaswal, H., *et al.* (2017). *Hybrid Storage Market Assessment: A JISEA White Paper.* Golden, CO, USA: National Renewable Energy Laboratory (NREL). https://doi.org/10.2172/1399357.

Eurek, K., Sullivan, P., Gleason, M., *et al.* (May 2017). An improved global wind resource estimate for integrated assessment models. *Energy Economics, 552–67.* https://doi.org/10.1016/j.eneco.2016.11.015.

Evans, M., Little, R., Lloyd, K., *et al.* (2007). Advanced modeling of renewable energy market dynamics. NREL/TP-640-41896. Available at https://www.nrel.gov/docs/fy07osti/41896.pdf.

Field, C. B., van Aalst, M. K., Adger, W. N., *et al.* (2014). *Part A: Global and Sectoral Aspects: Volume 1, Global and Sectoral Aspects: Working Group II Contribution to the Fifth Assessment Report of the Intergovernmental Panel on Climate Change.* Cambridge University Press, Cambridge, UK and New York, NY, USA.

Field, C. B., Barros, V. R., Mach, K. J., *et al.* (2014). Technical summary. In *Climate Change 2014: Impacts, Adaptation, and Vulnerability. Part A: Global and Sectoral Aspects. Contribution of Working Group II to the Fifth Assessment Report of the Intergovernmental Panel on Climate Change.* Cambridge University Press, Cambridge, UK and New York, NY, USA, pp. 35–94.

Field, C. B., *et al.* (2014). IPCC, 2014: Summary for policymakers. In *Climate Change 2014: Impacts, Adaptation, and Vulnerability. Part A: Global and Sectoral Aspects. Contribution of Working Group II to the Fifth Assessment Report of the Intergovernmental Panel on Climate Change.* Cambridge University Press, Cambridge, UK and New York, NY, USA, pp. 1–32.

Foust, T. D., Arent, D. J., Macedo, D. C. *et al.* (2015). Chapter 3: Energy security. In *Bioenergy and Sustainability: Bridging the Gaps.* NREL/CH-5100-66647.

Gernaat, D. E. H. J., Van Vuuren, D. P., Van Vliet, J., *et al.* (2014). global long-term cost dynamics of offshore wind electricity generation. *Energy,* 76, 663–672. https://doi.org/10.1016/j.energy.2014.08.062.

Hand, M. M., Baldwin, S., DeMeo, E., *et al.* (eds.). (2012). *Renewable Electricity Futures Study (Entire Report),* 4 vols. Golden, CO, USA: National Renewable Energy Laboratory (NREL), NREL/ TP-6A20-52409. Available at https://www.nrel.gov/analysis/re-futures.html.

Heath, G., Meldrum, J., Fisher, N., *et al.* (2014). Life cycle greenhouse gas emissions from barnett shale gas used to generate electricity. *Journal of Unconventional Oil and Gas Resources,* 8, 46–55. https://doi.org/10.1016/j.juogr.2014.07.002.

Heath, G. A., O'Donoughue, P., Arent, D. J., *et al.* (2014). Harmonization of initial estimates of shale gas life cycle greenhouse gas emissions for electric power generation. *Proceedings of the National Academy of Sciences,* 111(31), E3167–E3176. https://doi.org/10.1073/pnas.1309334111.

Hurlbut, D., Zhou, E., Porter, K., *et al.* (2015). *Renewables-Friendly Grid Development Strategies: Experience in the United States, Potential Lessons for China* (Chinese Translation). Golden, CO, USA: National Renewable Energy Labarotory (NREL). https://doi.org/10.2172/1278315.

Isley, S. C., Ketcham, R., and Arent, D. J. (2017). Using augmented reality to inform consumer choice and lower carbon footprints. *Environmental Research Letters*, 12(6), 064002. https://doi.org/10.1088/1748-9326/aa6def; https://www.nrel.gov/docs/fy17osti/65565.pdf.

Isley, S. C., Stern, P. C., Carmichael, S. P., *et al.* (2016). Online purchasing creates opportunities to lower the life cycle carbon footprints of consumer products. *Proceedings of the National Academy of Sciences*, 113(35), 9780–9785. https://doi.org/10.1073/pnas.1522211113; https://www.nrel.gov/docs/fy16 osti/65496.pdf.

Jenkin, T., Diakov, V., Drury, E., *et al.* (2013). *Use of Solar and Wind as a Physical Hedge against Price Variability within a Generation Portfolio*. Golden, CO, USA: National Renewable Energy Laboratory (NREL). https://doi. org/10.2172/1090959.

Johansson, T. B., Nakicenovic, N., Patwardhan, A., *et al.* (2012). Technical summary. Introduction. In *Global Energy Assessment: Toward a Sustainable Future* (pp. 31–94). Cambridge: Cambridge University Press. https://doi. org/10.1017/CBO9780511793677.006.

Kandt, A., Jun, J., Robinson, D. G., *et al.* (2008). *Risk-Based Approach To Incorporating Renewable Energy Into Critical Facilities*. Sandia National Laboratories, SAND2008-1101C.

Kocha, S.S., Peterson, M.W., and Arent, D. J. (1996). *Photoelectrochemical Based Direct Conversion Systems for Hydrogen Production*. Technical Report, USA. Available at https://www.osti.gov/servlets/purl/447148.

Kocha, S. S., Peterson, M. W., Arent, D. J., *et al.* (1995). Electrochemical investigation of the gallium nitride-aqueous electrolyte interface. *Journal of the Electrochemical Society*, 142(12), 238–240.

Kocha, S. S., Peterson, M. W., Nelson, A. J., *et al.* (1995). Investigation of chemical wet-etch surface modification of $Ga_{0.5}In_{0.5}P$ using photoluminescence, X-ray photoelectron spectroscopy, capacitance measurements, and photocurrent-voltage curves. *Journal of Physical Chemistry*, 99(2), 744–749. https://doi. org/10.1021/j100002a043.

Kwasnik, T., Carmichael, S. P., Arent, D. J., *et al.* (2017). *The Trip Itinerary Optimization Platform: A Framework for Personalized Travel Information.* Golden, CO, USA: National Renewable Energy Laboratory (NREL). https://doi.org/10.2172/1410410.

Lee, A., Zinaman, O., Logan, J., *et al.* (December 2012). Interactions, complementarities and tensions at the nexus of natural gas and renewable energy. *The Electricity Journal,* (10), 38–48. https://doi.org/10.1016/j.tej.2012.10.021.

Lee, N. and Arent, D. J. (2019). Introduction to the topical collection on regional renewable energy — Africa. *Current Sustainable/Renewable Energy Reports,* 6(1), 1–4. https://doi.org/10.1007/s40518-019-00124-5.

Logan, J. S., Arent, D. J., Zhou, S., *et al.* (2019). *Ten Principles for Power Sector Transformation in Emerging Economies.* Golden, CO, USA: National Renewable Energy Labarotory (NREL), NREL/TP-6A20-73931.

Logan, J., Lopez, A., Mai, T., *et al.* (2014). Natural gas scenarios in the US power sector. *Energy Economics,* 40, 183–195. https://doi.org/10.1016/j.eneco.2013.06.008.

Lowder, T., Schwabe, P., Zhou, E., *et al.* (2015). *Historical And Current U.S. Strategies For Boosting Distributed Generation* (Chinese Translation). Golden, CO, USA: National Renewable Energy Labarotory (NREL). https://doi.org/10.2172/1287332.

Luangdilok, C. H., Arent, D. J., Bocarsly, A. B., *et al.* (1992). Investigation of the structure-reactivity relationship in the platinum/metal cadmium hexacyanoferrate $(Pt/M_xCdFe(CN)_6)$-modified electrode system. *Langmuir,* 8(2), 650–657.

Luderer, G., Kriegler, E., Delsa, L., *et al.* (2016). *Deep Decarbonisation Towards 1.5°C–2°C Stabilisation.* Germany: Potsdam Institut Für Klimafolgenforschung.

Luderer, G., Pietzcker, R. C., Carrara, S., *et al.* (2017). Assessment of wind and solar power in global low-carbon energy scenarios: An introduction. *Energy Economics,* 64, 542–551. http://dx.doi.org/10.1016/j.eneco.2017.03.027.

Mai, T., Hand, M. M., Baldwin, S. F., *et al.* (2014). Renewable electricity futures for the United States. *IEEE Transactions on Sustainable Energy,* 5(2), 372–378. https://doi.org/10.1109/TSTE.2013.2290472.

Miller, M., Martinot, E., Cox, S., *et al.* (2015). *Power Systems of the Future: A 21st Century Power Partnership Thought Leadership Report.* Golden, CO,

USA: National Renewable Energy Laboratory (NREL), NREL/TP- 6A20-63366. Available at https://www.nrel.gov/docs/fy15osti/63278.pdf.

Miller, M., Martinot, E., Cox, S., *et al.* (2015). *Status Report on Power System Transformation*. Golden, CO, USA: National Renewable Energy Laboratory (NREL), NREL/TP6A2063366. Available at https://www.nrel.gov/docs/fy15osti/63366.pdf.

Milligan, M., Frew, B., Zhou, E., *et al.* (2015). *Advancing System Flexibility for High Penetration Renewable Integration* (Chinese Translation). Golden, CO, USA: National Renewable Energy Labarotory (NREL). https://doi.org/10.2172/1225920.

Moore, M. C., Arent, D. J., and Norland, D. (March 2007). R&D advancement, technology diffusion, and impact on evaluation of public R&D. *Energy Policy*, 35(3), 1464–1473. https://doi.org/10.1016/j.enpol.2006.03.009.

Muratori, M., Alexander, M., Arent, D. J., *et al.* (2020). The rise of electric vehicles — 2020 status and future expectations. *Progress in Energy*, 3(2), 022002. https://doi.org/10.1088/2516-1083/abe0ad.

Muratori, M., Jadun, P., Bush, B., *et al.* (2020). Future integrated mobility-energy systems: A modeling perspective. *Renewable and Sustainable Energy Reviews*, 119, 109541. https://doi.org/10.1016/j.rser.2019.109541.

Muratori, M., Jadun, P., Bush, B., *et al.* (2021). Exploring the future energy-mobility nexus: The transportation energy & mobility pathway options. TEMPO Model. *Transportation Research Part D: Transport and Environment*, 98, 102967. https://doi.org/10.1016/j.trd.2021.102967.

Nakicenovic, N., Grubler, A., Leininger, J., *et al.* (2020). *Innovations for Sustainability: Pathways to an Efficient and Post-Pandemic Future*. Laxenburg, Austria: International Institute for Applied Systems Analysis (IIASA), p. 108. doi: 10.22022/TNT/07-2020.16533.

Newmark, R., Arent, D. J., Sullivan, P., *et al.* (2010). Geospatial issues in energy-climate modeling: Implications for modelers, economists, climate scientists and policy makers. AGU Fall Meeting Abstracts.

Oakleaf, B., Cary, S., Meeker, D., *et al.* (2022). *A Roadmap toward a Sustainable Aviation Ecosystem*. Golden, CO, USA: National Renewable Energy Laboratory (NREL), NREL/TP-6A60-83060. Available at https://www.nrel.gov/docs/fy22osti/83060.pdf.

Paranhos, E., Kozak, T. G., Boyd, W., *et al.* (2015). *Controlling Methane Emissions in the Natural Gas Sector. A Review of Federal and State Regulatory Frameworks Governing Production, Gathering, Processing, Transmission, and Distribution.* Technical Report, USA. https://doi.org/10.2172/1215077.

Peterson, M. W., Turner, J. A., Parsons, C. A., *et al.* (1998). Miniband dispersion InGaAs/Al$_x$Ga$_{1-x}$As superlattices with wide wells and very thin barriers. *Applied Physics Letters,* 53, 2666–2668. https://doi.org/10.1063/1.100189.

Pless, J., Arent, D. J., Logan, J., *et al.* (2015). *Pathways to Decarbonization: Natural Gas and Renewable Energy: Lessons Learned from Energy System Stakeholders.* Golden, CO, USA: National Renewable Energy Laboratory (NREL), NREL/TP-6A50-63904. Available at https://doi.org/10.2172/1215173 and https://www.nrel.gov/docs/fy15osti/63904.pdf.

Pless, J., Arent, D. J., Logan, J., *et al.* (2016). Quantifying the value of investing in distributed natural gas and renewable electricity systems as complements: Applications of discounted cash flow and real options analysis with stochastic inputs. *Energy Policy,* 97, 378–390. https://doi.org/10.1016/j.enpol.2016.07.002.

Robinson, D. G., Arent, D. J., and Johnson, L. (2006). Impact of distributed energy resources on the reliability of critical telecommunications facilities. In *INTELEC 06 - Twenty-Eighth International Telecommunications Energy Conference.* 10–14 September 2006 Providence (USA), pp. 1–7. https://doi.org/10.1109/INTLEC.2006.251620.

Rogner, H.-H., Aguilera, R. F., Archer, C. L., *et al.* (2012). Energy resources and potentials. In *Global Energy Assessment: Toward a Sustainable Future* (pp. 425–512). Cambridge: Cambridge University Press. https://doi.org/10.1017/CBO9780511793677.

Romero-Lankao, P., Wilson, A., Sperling, J., *et al.* (2019). Urban electrification: Knowledge pathway toward an integrated research and development agenda. Mansueto Institute for Urban Innovation Research Paper No. 10. http://dx.doi.org/10.2139/ssrn.3440283.

Romero-Lankao, P., Wilson, A., Sperling, J., *et al.* (2022). Of actors, cities and energy systems: Advancing the transformative potential of urban electrification. *Progress In Energy,* 3(3). https://doi.org/10.1088/2516-1083/abfa25.

Rubin, H. D., Arent, D. J., and Bocarsly, A. B. (1985). The N-Cdse photo-electrochemical cell: Wavelength-dependent photostability. *Journal of the Electrochemical Society,* 132(2), 523.

Rubin, H. D., Arent, D. J., Humphrey, B. D., *et al.* (1987). Overlayer formation as a source of stability in the N-type photoelectrochemical cell. *Journal of Electrochemical Society,* 134, 93. https://doi.org/10.1149/1.2100444.

Sandor, D., Keyser, D., Mann, M., *et al.* (2020). Clean energy manufacturing: Renewable energy technology benchmarks. In *Accelerating the Transition to a 100% Renewable Energy Era* (pp. 195–206). Springer. https://doi.org/10.1007/978-3-030-40738-4.

Schaidle, J. A., Grim, R. G., Tao, L., *et al.* (2022). Power conversion technologies: The advent of power-to-gas, power-to-liquid, and power-to-heat. In G. Graditi and M. Di Somma (eds.), *Technologies for Integrated Energy Systems and Networks* (pp. 41–70). Wiley Online.

Schwabe, P., Mendelsohn, M., Mormann, F., *et al.* (2012). *Mobilizing Public Markets to Finance Renewable Energy Projects: Insights from Expert Stakeholders*. Golden, CO, USA: National Renewable Energy Laboratory (NREL), NREL/TP-6A20-55021. Available at https://www.nrel.gov/docs/fy12osti/55021.pdf.

Stark, C., Pless, J., Logan, J., *et al.* (2015). *Renewable Electricity: Insights for the Coming Decade*. Golden, CO, USA: National Renewable Energy Laboratory (NREL), NREL/TP-6A50-63604. https://doi.org/10.2172/1176740.

Taylor, R., Arent, D. J., Baldwin, S., *et al.* (1996). Opportunities and issues in international photovoltaic market development. In *Conference Record of the Twenty Fifth IEEE Photovoltaic Specialists Conference — 1996*, Washington, DC, USA, pp. 1465–1468. https://doi.org/10.1109/PVSC.1996.564412.

Turkenburg, W., Arent, D. J., Bertani, R., *et al.* (2012). Renewable energy. In *Global Energy Assessment: Toward a Sustainable Future* (pp. 761–900). Cambridge: Cambridge University Press. https://doi.org/10.1017/CBO9780511793677.

Turner, J. A., Arent, D. J., Peterson, M., *et al.* (1997). Photoelectrochemical systems for water decomposition. *Abstracts of Papers of the American Chemical Society,* 211.

Washington, W., Lee, K., Arent, D. J., *et al.* (2016). *Enhancing Participation in the U.S. Global Change Research Program*. USA. https://doi.org/10.17226/21837.

Zinaman, O., Miller, M., Adil, A., *et al.* (2015). Power systems of the future. *The Electricity Journal,* 28(2), 113–126. https://doi.org/10.1016/j.tej.2015.02.006.

Zinaman, O., Miller, M., Adil, A., *et al.* (2015). *Power Systems of the Future: A 21st Century Power Partnership Thought Leadership Report.* Golden, CO, USA: National Renewable Energy Laboratory (NREL), NREL/TP-6A20-62611. Available at https://doi.org/10.2172/1171135 and https://www.nrel.gov/docs/fy15osti/63278.pdf.

Zoback, M. D. and Arent, D. J. (2014). Shale gas: Development opportunities and challenges. *The Bridge,* 44(1). Available at https://www.nae.edu/111035/Shale-Gas-Development-Opportunities-and-Challenges#about_author 111035.

Zoback, M. D. and Arent, D. J. (August 2014). The opportunities and challenges of sustainable shale gas development. *Elements,* 10(4). https://doi.org/10.2113/gselements.10.4.251.

Appendix B
Solar Energy: Can It Be Used Effectively?

Douglas Arent
May, 25, 1976

This appendix provides a re-typeset of a paper the author wrote in 1976 that was the inspiration for his career in renewable energies, energy for sustainable development and sustainability more broadly. Photos and figures have not been reproduced here, but a PDF copy of the original document can be viewed in full at the following site: https://drive.google.com/file/d/1NH_FU22Y6nGFXWqyeQJitmLTjwRtcs8N/view?pli=1.

Table of Contents

Introduction

"We have reached the point where we must conserve, not expand, all our energy needs. We must also take a serious look at the different ways of generating electricity to see how they can be improved, and what the possibilities are for future use. Each of these methods has its own advantages and disadvantages. These must be considered when we choose our energy sources of the future."[1]

The earth's fossil fuels – coal, natural gas, oil, and oil shale, are diminishing extremely fast. With the demands for energy doubling every ten years, new and powerful sources of energy must be discovered and utilized effectively. Solar energy is one of these powerful sources.

Thermonuclear Reactions in the Sun

The sun's inner core temperature of 10 to 20 million degrees Centigrade or 2.9×10^9 degrees Fahrenheit allows the thermonuclear reactions to take place. At this temperature, hydrogen atoms lose their atomic composition (one proton attracted to one electron) dissociating into a free moving proton and a free moving electron.

With the kinetic energy (energy of motion) increasing directly with an increase in temperature and the sun having the extremely high inner core temperature mentioned above, the proton and electron move at a speed of 100 miles per second or 360,000 miles per hour. At this high speed, the hydrogen atoms are colliding an infinite number of times allowing for an infinite number of reactions (two substances must meet before they can react).

In these thermonuclear reactions, four hydrogen atoms combine to make one helium atom. Hydrogen has an atomic weight of 1.0077 grams, and when four hydrogen fuse into one helium, atomic weight 4.0, a loss of .0077 grams per gram of hydrogen is recorded. Since energy can neither be created or destroyed, the lost matter is converted into energy. These reactions are the sun's never ending source of power. (photo 1).

---

[1]Milton A. Rothman, *Energy and The Future* (New York: Franklin Watts Inc., 1975), p. 45.

Amount of Energy Released by the Sun

The sun releases an enormous amount of energy, 3.86×10^{33} ergs per second or 5.0×10^{23} horsepower, enough energy to power 5.0×10^{21} one hundred horsepower cars, 9.7×10^{7} times the number of cars in the United States today.

The earth receives only 1/2,200,000,000 of the sun's energy. In other words, 4.7 million horsepower per square mile or 1,350,000 ergs/cm^2/sec. Burning 5.5×10^{11} tons of coal would also produce this amount of energy, that is, 1,100,000,000,000,000 pounds of coal.

The United States receives 9,000 trillion kilowatt hours of the sun's energy annually, equivalent to the energy released by burning 1,150 billion tons of coal. This enormous amount of energy exceeds the needs of the country by a tremendous quantity. In 1969, if the U.S. utilized the sun's energy by using .14% of the land area (84 square miles or 1/1165 of Wyoming's total land area) at 10% efficiency, producers of energy, both public and private, wouldn't have used one gram of any of the fossil fuels.

How Long Will the Sun Last?

"The mass of the sun is 1.985×10^{33} grams, or, in energy terms, 1.785×10^{54} ergs. When we subtract the measured amount of annual radiation (1.4×10^{41} ergs), we are left with a value that, when divided by the approximately 6 million seconds in a year, indicates that the sun ought to last about 15 trillion years at its present rate of radiation."[2]

Another theory, based on the mass of hydrogen in the sun, estimates that if 5.64×10^{8} tons of hydrogen produce 5.6×10^{8} tons of helium and 4.0×10^{6} tons of energy, the sun will last only 30 billion years or 6.67×10^{9} generations of Man.

---

[2]Hans Rau, *Solar Energy* (New York: The MacMillan Co., 1964), p. 39.

Recent History

With the concern for Man's energy needs emerging during the early 1900's, when projections were made about the depleting fossil fuels, the interest in the sun as an energy source has been hindered by the availability and relatively easy development of the fossil fuels. Recently, many new and feasible methods of harnessing the sun's energy have been invented.

In 1918, E.E. Willise and J. Boyle invented the solar heater by using two circulating liquids. The heat was absorbed by a flat, glass covered water tank and then transported to liquid sulfur dioxide. The heated sulfur dioxide volatilized (changing from liquid form to a vaporous phase). The vapors were then used to drive a steam engine.

Since the engine only produced fifteen horsepower at its maximum, it was never a great prospect as a solar energy converter.

Charles G. Abbot, in 1930, invented a solar boiler. A 300 square foot reflector focused the sun's rays onto a black tube which was enclosed in double-wall glass insulation. A liquid which was circulated through the black tube was passed to the cooker by gravitational flow and the Laws of Thermal Expansion.

In 1954, the Bell Telephone Laboratories invented the solar battery. The battery consists of 432 solar cells, each covering an area of about one square inch. The cells are sealed in plastic and covered by a layer of glass. These cells convert sunlight directly into electricity. (photo 2).

The research of the Bell Laboratories was not extremely important in 1954, but now, with the earth's fossil fuels diminishing at a drastic rate, producers and consumers of energy are becoming increasingly concerned with the Bell experiments as the foundation for meeting the current and foreseeable future energy needs of the world.

Collection and Concentration

"Solar energy is very dilute. It is spread out over a large area, so that it must be collected and concentrated in order to be useful."[3]

---

[3]Milton A. Rothman, *Energy and The Future* (Franklin Watts Inc., 1975), p. 53.

The collection and concentration is done by two basic methods. In one method, parabolic mirrors focus the sun's rays on a black surface, such as a tube. The intensity of these pinpoint rays produces temperatures up to 6,000 degrees Centigrade or 10,800 degrees Fahrenheit, requiring the tube to be enclosed in two or three layers of glass to lessen the heat loss.

The liquid water which is circulated through the tube vaporizes. The vapor then activates a turbine that produces power. The mirrors themselves are made of glass, aluminum or aluminized plastic and are rotated to keep facing the sun.

The second method utilizes flat plate collectors. A black metal sheet is covered with three layers of protecting glass with air added between as an extra insulator to reduce heat loss. The hot air circulates through tubes and is then used to either heat houses or operate engines.

The solar batteries developed by the Bell Labs in 1954 convert sunlight directly into electricity. A positive and a negative electrode are enclosed in silicon, a semi-conductor of electricity. When sunlight hits the silicon, its conductivity increases and an electrical circuit is completed.

The energy produced by a single cell is small, but when many cells are combined, they can conduct a substantial amount of electricity – enough to run a phone or a small household appliance, such as, a toaster or a hair dryer.

Current Usages

The ability of the solar furnace to produce temperatures as high as 8,500 degrees Centigrade make it an excellent industrial tool. The heat, provided by parabolic mirrors, is used to test metals and to bake ceramics. (photo 3).

The solar engine harnesses the sun's energy directly, without transferring it into water, sulfur dioxide or any other substance. The engine has a double layer of glass on top with air inside the cylinder. When the sunlight hits the top, the heat expands the air inside forcing the piston down. The expanded air cools and the piston reverses the action by compressing the air. The process repeats itself producing power. (photo 4).

Solar electricity is a major concern now. As mentioned above, the process of converting solar energy into electricity was developed by the Bell Laboratories. The methods for producing solar electricity have been established and proven, but are not feasible at this time for widespread use because of the prohibiting production costs involved.

Solar houses, one of the more important affairs presently, use flat plate collectors or solar batteries. These houses produce up to 80% of their own energy needs. (photos 5 and 6).

In 1958, the Massachusetts Institute of Technology completed a solar house at Lexington Mass. The house uses a 640 square foot flat plate collector. The heated air is forced through ducts to heat the house, or passed to tanks of chemical salts where the heat content of the air can be transferred to the salts for storage. Professor Dietz, the supervisor of the project, said "The experiment's principal value lies in the knowledge gained in constructing and operating a complex solar heating plant. Eventually, I believe, the sun's energy will be used widely as a fuel for heating systems. Meanwhile, solar energy in countries having abundant conventional fuels will be economically feasible only when the climate provides an unusually high yield of sunshine."[4]

Another experimental house was developed in Delaware at the University of Delaware of Energy Conversion at Newark. Solar energy from the solar cells on the roof provide for 80% of the total electricity needs. The energy which is not used immediately is stored in lead-acid batteries for night and cloudy day use.

The heating system consists of hot air being circulated through the house during the day. While the air still contains a substantial amount of heat, it is transported into containers filled with chemical salts. Gauber's salts, as they are called, are carefully prepared to have a melting point between 90 and 100 degrees Fahrenheit. This quality allows for the least amount of energy to be wasted when the salts are used to store heat.

---

[4]Hans Rau, *Solar Energy* (New York: The MacMillan Co., 1964), p. 64.

When the heated air comes in contact with the salts, the salts melt, consequently storing heat energy for night and cold day use. The salts can store heat energy up to ten days.

In the summer, the system works in reverse. The cool night air freezes the salts. Then during the day, when cool air is needed to lower the temperature of the house, the salts are allowed to melt, giving off the required cool air.

Much research is being put into solar houses. Many new types of houses and heating systems are being developed using solar batteries, parabolic mirrors, and the flat plate collectors. Even laws have been passed to help the solar house owner get the needed sunlight to run his house. In most instances, the laws affect local zoning and the height of neighboring trees.

To harness an adequate amount of the sun's energy, a large quantity of land must be used. To produce the normal 15,000 megawatts of electricity that an average power plant produces, 10 square miles would have to be covered with solar batteries. Productive, open land area is becoming more critical with the global food shortage; and therefore, science cannot further its large scale research since the study requires the land.

Future

Scientists know that the sun's energy can be effectively used – workable engines and solar houses have been demonstrated. But because most projections for large scale use are based on hypothetical theories, anticipated fuel and material costs, and predicted population growth, more research must be completed, and techniques and equipment refined in order to achieve an effective but economical method to utilize solar energy.

One plan for large scale production consists of placing a huge collector in orbit around the earth. The collector will then transmit energy waves to the earth where they can be successfully used. This space system is environmentally satisfactory, but the cost of undertaking the launch of a solar collector would be enormous — too large for present day investment and technology (photo 7).

In upcoming years, when the earth's fossil fuels are defunct, this plan for launching a collector into orbit will be an extremely small task.

"It is still too early to make accurate assessments of the economic feasibility of solar energy, but the prospects appear to be encouraging enough to warrant further research. Similar economic questions remain to be resolved about breeder reactors, and fusion power systems have yet to be proved scientifically feasible, so that solar energy must be considered a significant, if still uncertain, alternative for future power needs."[5]

---

[5]Allen L. Hammond, "Solar Energy: A Feasible Source of Power", *Science* (May, 14, 1971), p. 660.

Table of Definitions

Calorie – the amount of energy needed to raise the temperature of one gram of water one degree Centigrade.

Electron – the negatively charged particle of an atom.

Egrs – a unit of measure for energy. 4.186×10^7ergs = one calorie.

Neutron – a neutral particle of an atom.

Nucleus – the center of an atom.

Proton – a positively charged particle of an atom which is located in the nucleus.

Scientific Notation – a way of expressing numerals including a power of ten.
Ex. $10^2 = 100$ $1.632 \times 10^8 = 163,200,000$

Bibliography

Branley, Franklyn M. *The Sun, Star Number One*. New York: Thomas Y. Crowell Comp., 1964

Hammond, Allen L. "Solar Energy: A Feasible Source of Power", *Science* (May, 14, 1971), p. 660

Henahan, John F. "How Soon the Sun: A Timetable for Solar Energy", *Saturday World Review* (November, 20,1973), pp. 46–66

Rau, Hans. *Solar Energy*. New York: The MacMillan Comp., 1964

Rothman, Milton A. *Energy and the Future*. New York: Franklin Watts Inc., 1975

"Solar Energy", *The World Book Encyclopedia* (11th Ed.), Vol. 15, 7546a–7547

References

1. Available at https://www.energy.gov/sites/default/files/ERDA%20History. pdf.

2. Becquerel, E. (1839). Available at https://www.smithsonianmag.com/ sponsored/brief-history-solar-panels-180972006/.

3. Energy strategy: The road not taken? — RMI. (n.d.) RMI. Available at https://www.facebook.com/RockyMtnInst/ (Accessed August 13 2023). Available at https://rmi.org/insight/energy-strategy-the-road-not-taken/.

4. Nakicneovic, N. (1997). Technological change as a learning process. International Institute of Applied Systems Analysis. Available at https:// pure.iiasa.ac.at/id/eprint/12530/1/Technological%20Change%20as%20 a%20Learning%20Process.pdf.

5. Nemet, G. F. (November 2006). Beyond the learning curve: Factors influencing cost reductions in photovoltaics. *Energy Policy*, (17), 3218–32. https://doi.org/10.1016/j.enpol.2005.06.020.

6. Singh, M., and Santoso, S. (2011). Dynamic models for wind turbines and wind power plants. NREL/SR-5500-52780. Available at https://www.nrel. gov/docs/fy12osti/52780.pdf.

7. Veers, P., Dykes, K., Basu, S., Bianchini, A., Clifton, A., Green, P., Holttinen, H., *et al.* (December 2022). Grand challenges: Wind energy research needs for a global energy transition. *Wind Energy Science*, (6), 2491–96. https://doi.org/10.5194/wes-7-2491-2022.

8. Bauer, J. The leading edge: Wind energy newsletter, illustration. NREL, 82082. Available at https://www.nrel.gov/wind/newsletter-202004.html.

9. Veers, P., *et al.* (2019). Grand challenges in the science of wind energy. *Science*, 366(6464), eaau2027. https://doi.org/10.1126/science.aau2027.

10. *Photovoltaics Report.* (November 4 2016). Learning curve based on EuPD data (Fraunhofer ISE October 2016). Available at https://

www.connaissancedesenergies.org/sites/default/files/pdf-actualites/
photovoltaics-report_fraunhofer_2016.pdf.

11. Available at https://www.eia.gov/todayinenergy/detail.php?id=8870 and
An analysis of the costs, benefits, and implications of different approaches
to capturing the value of renewable energy tax incentives, Bolinger,
Mark, LBNL, Available at https://escholarship.org/content/qt40n5f87b/
qt40n5f87b.pdf.

12. Parinandi, S. C. (2020). Policy inventing and borrowing among state
legislatures. *American Journal of Political Science.* https://doi.org/10.1111/
ajps.12513.

13. IEA. (2013). Electricity feed-in law of 1991 ("Stromeinspeisungsgesetz").
Available at https://www.iea.org/policies/3477-electricity-feed-in-law-of-
1991-stromeinspeisungsgesetz.

14. Couture, T., Cory, K., Kreycik, C., and Williams, E. (2010). *A Policymaker's
Guide to Feed-in Tariff Policy Design.* Golden, CO, USA: National
Renewable Energy Laboratory (NREL), NREL/TP-6A2-44849. Available
at https://www.nrel.gov/docs/fy10osti/44849.pdf.

15. European Commission. (1997). A strategy and action plan for the
promotion of renewable energy sources. Available at https://ec.europa.eu/
commission/presscorner/detail/en/IP_97_1040.

16. IEA. (2017). Feed-in tariffs for electricity from renewable energy sources
(Special regime). Available at https://www.iea.org/policies/4555-feed-in-
tariffs-for-electricity-from-renewable-energy-sources-special-regime.

17. Ackermann, T. and Soeder, L. (2002). An overview of wind energy-status
2002. *Renewable and Sustainable Energy Reviews,* 6, 67–128.

18. Turner, J. A. (July 1999). A realizable renewable energy future. *Science,*
(5428), 687–89. https://doi.org/10.1126/science.285.5428.687.

19. IRENA. (2022). Renewable capacity statistics 2022. IRENA. Available at
https://www.irena.org/publications/2022/Apr/Renewable-Capacity-
Statistics-2022.

20. EIA. (2023). What is the difference between electricity generation capacity
and electricity generation? Available at https://www.eia.gov/tools/faqs/faq.
php?id=101&t=3.

21. Wilson, C. and Tyfield, D. (2018). Critical perspectives on disruptive
innovation and energy transformation. *Energy Research & Social Science,* 37,

211–215. https://doi.org/10.1016/j.erss.2017.10.032, and the full Special Issue.

22. Arndt, C., Arent, D., Hartley, F., Merven, B., and Mondal, A. H. (October 2019). Faster than you think: Renewable energy and developing countries. *Annual Review of Resource Economics*, (1), 149–68. https://doi.org/10.1146/annurev-resource-100518-093759.

23. Grubler, A. (2010). The costs of the French nuclear scale-up: A case of negative learning by doing. *Energy Policy,* 38, 5174–5188.

24. Rennert, K., Prest, B. C., Pizer, W. A., Newell, R. G., Anthoff, D., Kingdon, C., Rennels, L., Cooke, R., Raftery, A. E., Ševčíková, H., and Errickson, F. (2022). The social cost of carbon: Advances in long-term probabilistic projections of population, GDP, emissions, and discount rates. *Brookings Papers on Economic Activity*, 2021(2), 223–305. Brookings Institution Press. Available at https://doi.org/10.1353/eca.2022.0003.

25. EIA. (2022). What is the efficiency of different types of power plants? Available at https://www.eia.gov/tools/faqs/faq.php?id=107&t=3.

26. Neuhoff, K., May, N., and Richstein, J. C. (2022). Financing renewables in the age of falling technology costs. *Resource and Energy Economics*, 70, 101330. https://doi.org/10.1016/j.reseneeco.2022.101330.

27. Stiglitz, J. (2020). GDP is the wrong tool for measuring what matters: It's time to replace gross domestic product with real metrics of well-being and sustainability. *Scientific American*, 323(2). https://www.scientificamerican.com/article/gdp-is-the-wrong-tool-for-measuring-what-matters/.

28. Sipiczki, A. (2022). A critical look at the ESG market. CEPS Policy Insights No. 2022-15. Available at https://www.ceps.eu/wp-content/uploads/2022/04/PI2022-15_A-critical-look-at-the-ESG-market.pdf.

29. Boffo, R. and Patalano, R. (2020). *ESG Investing: Practices, Progress and Challenges*. Paris: OECD. Available at https://www.oecd.org/finance/ESG-Investing-Practices-Progress-Challenges.pdf.

30. For a detailed calculation of full LCOE and influencing factors. Available at https://atb.nrel.gov/electricity/2022/definitions; https://atb.nrel.gov/electricity/2022/equations_&_variables.

31. Wang, Z. and Krupnick, A. (2013). US shale gas development what led to the boom? *Resources for the Future*. Available at https://media.rff.org/documents/RFF-IB-13-04.pdf.

32. Arent, D., Logan, J., Macknick, J., *et al.* (2015). A review of water and greenhouse gas impacts of unconventional natural gas development in the United States. *MRS Energy & Sustainability*, 2, 4. https://doi.org/10.1557/mre.2015.5.

33. Pless, J., Arent, D. J., Logan, J., Cochran, J., and Zinaman, O. (2016). Quantifying the value of investing in distributed natural gas and renewable electricity systems as complements: Applications of discounted cash flow and real options analysis with stochastic inputs. *Energy Policy*, 97, 378–390. https://doi.org/10.1016/j.enpol.2016.07.002.

34. Huang, P., Negro, S. O., Hekkert, M. P., and Bi, K. (October 2016). How China became a leader in solar PV: An innovation system analysis. *Renewable and Sustainable Energy Reviews*, 777–789. https://doi.org/10.1016/j.rser.2016.06.061.

35. REN21. (2022). *Renewables 2021 Global Status Report*. Available at https://www.ren21.net/wp-content/uploads/2019/05/GSR2021_Full_Report.pdf.

36. Davis, S. J., Lewis, N. S., Shaner, M., Aggarwal, S., Arent, D., Azevedo, I. L., Benson, S. M., *et al.* (June 2018). Net-zero emissions energy systems. *Science*, (6396). https://doi.org/10.1126/science.aas9793.

37. Azevedo, I., Bataille, C., Bistline, J., Clarke, L., and Davis, S. (December 2021). Introduction to the special issue on net-zero energy systems. *Energy and Climate Change*, 100066. https://doi.org/10.1016/j.egycc.2021.100066.

38. O'Shaughnessy, E., Heeter, J., Shah, C., and Koebrich, S. (2021). Corporate acceleration of the renewable energy transition and implications for electric grids. *Renewable and Sustainable Energy Reviews*, 146, 111160. https://doi.org/10.1016/j.rser.2021.111160.

39. IRENA. (2019). Renewable energy auctions: Status and trends beyond price. IRENA. Available at https://www.irena.org/-/media/Files/IRENA/Agency/Publication/2019/Jun/IRENA_Auctions_beyond_price_2019_findings.pdf.

40. European Commission. (2018). Renewable energy directive. Available at https://energy.ec.europa.eu/topics/renewable-energy/renewable-energy-directive-targets-and-rules/renewable-energy-directive_en.

41. Vidal, A. D. (2022). Cloudy days for solar energy: A study of the Spanish solar photovoltaic feed-in tariff through the multi-level perspective and socio-technical transition pathways. *Intersect: The Stanford Journal of Science, Technology, and Society*, 16(1).

42. Kroposki, B., *et al.* (March–April 2017). Achieving a 100% renewable grid: Operating electric power systems with extremely high levels of variable renewable energy. *IEEE Power and Energy Magazine*, 15(2), 61–73. https://doi.org/10.1109/MPE.2016.2637122.

43. U.S. Department of Energy. (2008). 20% Wind energy by 2030 increasing wind energy's contribution to U.S. electricity supply. Available at https://www.nrel.gov/docs/fy08osti/41869.pdf.

44. Short, W., Blair, N., and Heimiller, D. (November/December 28 2003). The long-term potential of wind power in the United States. *Solar Today*. Available at https://www.nrel.gov/docs/gen/fy04/34871.pdf.

45. Hand, M. M., Baldwin, S., DeMeo, E., Reilly, J. M., Mai, T., Arent, D., Porro, G., Meshek, M., and Sandor, D. (eds.). (2012). *Renewable Electricity Futures Study (Entire Report)*, 4 vols. Golden, CO, USA: National Renewable Energy Laboratory (NREL), NREL/TP-6A20-52409. Available at https://www.nrel.gov/analysis/re-futures.html.

46. Lopez, A., Roberts, B., Heimiller, D., Blair, N., and Porro, G. (2012). *U.S. Renewable Energy Technical Potentials: A GIS-Based Analysis*. Golden, CO, USA: National Renewable Energy Laboratory (NREL). Available at https://www.nrel.gov/docs/fy12osti/51946.pdf.

47. Ho, J., Becker, J., Brown, M., Brown, P., Chernyakhovskiy, I., Cohen, S., Cole, W., *et al.* (2021). *Regional Energy Deployment System (ReEDS) Model Documentation: Version 2020*. Golden, CO, USA: National Renewable Energy Laboratory (NREL), NREL/TP-6A20-78195. Available at https://www.nrel.gov/analysis/reeds/.

48. Cochran, J., Mai, T., Bazilian, M. (2014). Meta-analysis of high penetration renewable energy scenarios. *Renewable and Sustainable Energy Reviews, 29*, 246–53. https://doi.org/10.1016/j.rser.2013.08.089.

49. Joskow, P. L. and Schmalensee, R. (December 1988). *Markets for Power: An Analysis of Electrical Utility Deregulation* (1st edn.), Vol. 1. MIT Press Books, no. 0262600188.

50. Borenstein, S. and Bushnell, J. (2015). The US electricity industry after 20 years of restructuring. *Annual Review of Economics, 7*, 437–463. https://doi.org/10.1146/annurev-economics-080614-115630.

51. Cochran, J., Lew, D., and Kumar, N. (2014). Making coal flexible: Getting from baseload to peaking plant. *Cornerstone*, 2(4), 41–45. Available at https://bsm.nrel.gov/assets/pdfs/cornerstone-flexible-coal.pdf.

52. Kroposki, B. (2018). Integrating high levels of variable renewable energy into electric power systems. NREL/PR-5D00-68349. Available at https://www.nrel.gov/docs/fy17osti/68349.pdf.

53. IEA. (2019). Status of power system transformation 2019: Power system flexibility. *IEA and 21st Century Power Partnership Report.* https://iea.blob.core.windows.net/assets/00dd2818-65f1-426c-8756-9cc0409d89a8/Status_of_Power_System_Transformation_2019.pdf.

54. Cochran, J., Bird, L., Heeter, J., and Arent, D. J. (April 2012). *Integrating Variable Renewable Energy in Electric Power Markets: Best Practices from International Experience.* Golden, CO, USA: National Renewable Energy Laboratory (NREL), NREL/TP-6A00-53732. Available at https://doi.org/10.2172/1041369 and https://www.nrel.gov/docs/fy12osti/53732.pdf.

55. Johansen, K. (May–June 2021). Wind energy in Denmark: A short history. *IEEE Power and Energy Magazine*, 19(3), 94–102. https://doi.org/10.1109/MPE.2021.3057973.

56. Loutan, C., Klauer, P., Chowdhury, S., Hall, S., Morjaria, M., Chadliev, V., Milam, N., Milan, C., and Gevorgian, V. (2017). *Demonstration of Essential Reliability Services by a 300-MW Solar Photovoltaic Power Plant.* Golden, CO, USA: National Renewable Energy Laboratory (NREL), NREL/TP-5D00-67799. Available at https://doi.org/10.2172/1349211.

57. Chernyakhovskiy, I., Koebrich, S., Gevorgian, V., and Cochran, J. (2019). *Grid-Friendly Renewable Energy: Solar and Wind Participation in Automatic Generation Control Systems.* Golden, CO, USA: National Renewable Energy Laboratory (NREL), NREL/TP-6A20-73866. Available at https://doi.org/10.2172/1543130 and https://www.nrel.gov/docs/fy19osti/73866.pdf.

58. Wang, Q., Martinez-Anido, C. B., Wu, H., Florita, A. R., and Hodge, B.-M. (October 16 2016). Quantifying the economic and grid reliability impacts of improved wind power forecasting. *IEEE Transactions on Sustainable Energy*, 7(4), 1525–37. https://doi.org/10.1109/TSTE.2016.2560628.

59. Energy Modeling Forum. Available at https://emf.stanford.edu/.

60. IPCC, 2018: Global Warming of 1.5°C. *An IPCC Special Report on the Impacts of Global Warming of 1.5°C Above Pre-industrial Levels and Related Global Greenhouse Gas Emission Pathways, in the Context of Strengthening the Global Response to the Threat of Climate Change, Sustainable Development,*

and Efforts to Eradicate Poverty [V. Masson-Delmotte, P. Zhai, H.-O. Pörtner, D. Roberts, J. Skea, P. R. Shukla, A. Pirani, W. Moufouma-Okia, C. Péan, R. Pidcock, S. Connors, J. B. R. Matthews, Y. Chen, X. Zhou, M. I. Gomis, E. Lonnoy, T. Maycock, M. Tignor, and T. Waterfield (eds.)]. Cambridge University Press, Cambridge, UK and New York, NY, USA, 616 pp. https://doi.org/10.1017/9781009157940.

61. IPCC, 2023: Sections. In: *Climate Change 2023: Synthesis Report. Contribution of Working Groups I, II and III to the Sixth Assessment Report of the Intergovernmental Panel on Climate Change* [Core Writing Team, H. Lee and J. Romero (eds.)]. IPCC, Geneva, Switzerland, pp. 35–115, https://www.ipcc.ch/report/ar6/syr/downloads/report/IPCC_AR6_SYR_FullVolume.pdf and details of mitigation pathways and scenarios in IPCC, 2022. Summary for policymakers. [P. R. Shukla, J. Skea, A. Reisinger, R. Slade, R. Fradera, M. Pathak, A. Al Khourdajie, M. Belkacemi, R. van Diemen, A. Hasija, G. Lisboa, S. Luz, J. Malley, D. McCollum, S. Some, and P. Vyas (eds.)] In: *Climate Change 2022: Mitigation of Climate Change. Contribution of Working Group III to the Sixth Assessment Report of the Intergovernmental Panel on Climate Change* [P. R. Shukla, J. Skea, R. Slade, A. Al Khourdajie, R. van Diemen, D. McCollum, M. Pathak, S. Some, P. Vyas, R. Fradera, M. Belkacemi, A. Hasija, G. Lisboa, S. Luz, and J. Malley (eds.)]. Cambridge University Press, Cambridge, UK and New York, NY, USA. https://doi.org/10.1017/9781009157926.001.

62. Larson, E., Greig, C., Jenkins, J., Mayfield, E., Pascale, A., Zhang, C., Drossman, J., Williams, R., Pacala, S., Socolow, R., Baik, E. J., Birdsey, R., Duke, R., Jones, R., Haley, B., Leslie, E., Paustian, K., and Swan, A. (October 29 2021). *Net-Zero America: Potential Pathways, Infrastructure, and Impacts*, Final Report Summary. Princeton, NJ: Princeton University. https://netzeroamerica.princeton.edu/the-report.

63. Denholm, P., Brown, P., Cole, W., *et al.* (2022). *Examining Supply-Side Options to Achieve 100% Clean Electricity by 2035*. Golden, CO, USA: National Renewable Energy Laboratory (NREL), NREL/TP6A40-81644. Available at https://www.nrel.gov/docs/fy22osti/81644.pdf and https://www.nrel.gov/news/program/2022/exploring-the-big-challenge-ahead-insights-on-the-path-to-a-net-zero-power-sector-by-2035.html.

64. Joskow, P. L. Introduction to electricity sector liberalization: Lessons learned from cross country studies. In F. Sioshansi and W. Pfaffenberger (eds.), *Electricity Market Reform: An International Perspective*, pp. 1–32.

65. Hurlbut, D. (2017). *Creative Destruction and the Electric Utility of the Future.*

66. Milligan, M., Frew, B., Zhou, E., and Arent, D. J. (2015). Advancing system flexibility for high penetration renewable integration. USA. https://doi.org/10.2172/1225920 and https://www.osti.gov/servlets/purl/1225920.

67. Bowen, T., Chernyakhovskiy, I., and Denholm, P. L. (2019). Grid-scale battery storage: Frequently asked questions. USA. https://doi.org/10.2172/1561843 and https://www.osti.gov/servlets/purl/1561843.

68. Blair, N., Augustine, C., Cole, W., Denholm, P., Frazier, W., Geocaris, M., Jorgenson, J., McCabe, K., Podkaminer, K., Prasanna, A., and Sigrin, B. (2022). *Storage Futures Study: Key Learnings for the Coming Decades.* Golden, CO, USA: National Renewable Energy Laboratory (NREL), NREL/TP-7A40-81779. Available at https://doi.org/10.2172/1863547 and https://www.osti.gov/servlets/purl/1863547.

69. Storage Futures | Energy Analysis | NREL. (n.d.). National Renewable Energy Laboratory (NREL) home page. NREL. https://www.nrel.gov/analysis/storage-futures.html (Accessed August 19 2023).

70. Jenkins, J. D. and Sepulveda, N. A. (September 2021). Long-duration energy storage: A blueprint for research and innovation. *Joule*, (9), 2241–46. https://doi.org/10.1016/j.joule.2021.08.002.

71. Bistline, J., Cole, W., Damato, G., DeCarolis, J., Frazier, W., Linga, V., Marcy, C., Namovicz, C., Podkaminer, K., Sims, R., Sukunta, M., and Young, D. (2020). Energy storage in long-term system models: A review of considerations, best practices, and research needs. UK. https://doi.org/10.1088/2516-1083/ab9894.

72. Sovacool, B. K. (2018). Success and failure in the political economy of solar electrification: Lessons from World Bank Solar Home System (SHS) projects in Sri Lanka and Indonesia. *Energy Policy*, 123(C), 482–493.

73. HOMER Energy. (n.d.). Available at https://www.homerenergy.com/.

74. Renewable energy optimization model. Available at https://reopt.nrel.gov/.

75. Denholm, P., King, J. C., Kutcher, C. F., and Wilson, P. P. H. (2012). Decarbonizing the electric sector: Combining renewable and nuclear energy using thermal storage. *Energy Policy*, 44(C), 301–311. https://doi.org/10.1016/j.enpol.2012.01.055.

76. Arent, D. J., Bragg-Sitton, S. M., Miller, D. C., Tarka, T. J., Engel-Cox, J. A., Boardman, R. D., Balash, P. C., Ruth, M. F., Cox, J., and Garfield, D. J. (2021). Multi-input, multi-output hybrid energy systems. USA. https://doi.org/10.1016/j.joule.2020.11.004.

77. U.S. Department of Energy. (2023). *U.S. National Clean Hydrogen Strategy and Roadmap*. Available at https://www.hydrogen.energy.gov/pdfs/us-national-clean-hydrogen-strategy-roadmap.pdf.

78. European Union. EU strategy on energy system integration. https://energy.ec.europa.eu/topics/energy-systems-integration/eu-strategy-energy-system-integration_en (Accessed August 2023).

79. Available at https://www.seia.org/initiatives/solar-investment-tax-credit-itc.

80. Available at https://eur-lex.europa.eu/legal-content/EN/ALL/?uri=CELEX%3A32009L0028.

81. Available at https://eur-lex.europa.eu/legal-content/EN/TXT/?uri=uriserv%3AOJ.L_.2001.283.01.0033.01.ENG&toc=OJ%3AL%3A2001%3A283%3ATOC.

82. Available at https://eur-lex.europa.eu/legal-content/EN/TXT/?uri=uriserv%3AOJ.L_.2003.123.01.0042.01.ENG&toc=OJ%3AL%3A2003%3A123%3ATOC.

83. Available at https://www.solarpowerworldonline.com/2015/09/u-s-solar-industry-surpasses-20-gw-of-operational-solar-capacity/.

84. Available at https://www.youtube.com/watch?v=E7yABGDD85Q.

85. Available at https://www.solarpowerworldonline.com/2017/02/u-s-solar-market-sees-astounding-95-growth-2016/.

86. Available at https://www.solarpowerworldonline.com/2017/09/nrel-report-shows-utility-scale-solar-costs-declined-30-since-last-year/.

87. United Nations. (n.d.). Nationally determined contributions registry. Available at https://unfccc.int/NDCREG (Accessed August 2023).

88. International Energy Agency. (2018). *The Future of Cooling*. Available at https://www.iea.org/reports/the-future-of-cooling.

89. Autonomous Energy Systems | Grid Modernization | NREL. (n.d.). National Renewable Energy Laboratory (NREL) Home Page | NREL. Available at https://www.nrel.gov/grid/autonomous-energy.html (Accessed August 19 2023).

90. Lundquist, J. K., DuVivier, K. K., Kaffine, D., and Tomaszewski, J. M. M. Costs and consequences of wind turbine wake effects arising from uncoordinated wind energy development. United States. https://doi.org/10.1038/s41560-018-0281-2 and https://www.osti.gov/servlets/purl/1484339.

91. Allen, J. M., King, R. N., and Barter, G. E. (2020). Wind farm simulation and layout optimization in complex Terrain: Preprint. United States. Available at https://doi.org/10.1088/1742-6596/1452/1/012066 and https://www.osti.gov/servlets/purl/1603879.

92. ExaWind – Exascale Computing Project. (n.d.). Exascale computing project. Available at https://www.exascaleproject.org/research-project/exawind/ (Accessed August 19 2023).

93. Bernstein, A. (2022). *Real-Time Optimization and Control of Autonomous Energy Systems: From Theory to Practice.* USA: National Renewable Energy Laboratory (NREL), NREL/PR-5D00-82922. Available at https://www.nrel.gov/grid/assets/pdfs/aes-sp3-algorithms.pdf and https://www.nrel.gov/docs/fy22osti/82922.pdf.

94. Matevosyan, J., *et al.* (November–December 2021). A future with inverter-based resources: Finding strength from traditional weakness. *IEEE Power and Energy Magazine*, 19(6), 18–28. https://doi.org/10.1109/MPE.2021.3104075.

95. Isley, S. C., Stern, P. C., Carmichael, S. P., Joseph, K. M., and Arent, D. J. (2016). Online purchasing creates opportunities to lower the life cycle carbon footprints of consumer products. *Proceedings of the National Academy of Sciences*, 113(35), 9780–9785.

96. Montmasson-Clair, G. and Ryan, G. (2014). Lessons from South Africa's renewable energy regulatory and procurement experience. *Journal of Economic and Financial Sciences*, 7(si-1).

97. There is a significant literature on this today. Willet Kempton championed this idea nearly two decades ago. See for example, Kempton, W. and Tomić, J. (2005). Vehicle-to-grid power fundamentals: Calculating capacity

and net revenue. *Journal of Power Sources*, 144(1), 268–279. https://doi. org/10.1016/j.jpowsour.2004.12.025.

98. IEA. (2020). *Austria 2020*. Paris: IEA. Available at https://www.iea.org/ reports/austria-2020.

99. White House Briefing. (2021). FACT SHEET: President Biden sets 2030 greenhouse gas pollution reduction target aimed at creating good-paying union jobs and securing U.S. Leadership on Clean Energy Technologies. Available at https://www.whitehouse.gov/briefing-room/statements-releases/ 2021/04/22/fact-sheet-president-biden-sets-2030-greenhouse-gas-pollution-reduction-target-aimed-at-creating-good-paying-union-jobs-and-securing-u-s-leadership-on-clean-energy-technologies/.

100. Larsen, J., King, B., Kolus, H., and Herndon, W. (2021). Pathways to build back better: Investing in 100% clean electricity. Rhodium Group. Available at https://rhg.com/wp-content/uploads/2021/03/Pathways-to-Build-Back-Better-Investing-in-100-Clean-Electricity.pdf.

101. Denholm, P., Brown, P., Cole, W., *et al.* (2022). *Examining Supply-Side Options to Achieve 100% Clean Electricity by 2035*. Golden, CO, USA: National Renewable Energy Laboratory (NREL), NREL/TP6A40-81644. Available at https://www.nrel.gov/docs/fy22osti/81644.pdf.

102. Electric Power Research Institute. (2021). Powering decarbonization strategies for net-zero CO_2 emissions. EPRI. Available at https://www.epri. com/research/programs/109396/results/3002020700.

103. IEA. (2022). *World Energy Outlook 2022*. Paris: IEA. Available at https:// www.iea.org/reports/world-energy-outlook-2022, License: CC BY 4.0 (report); CC BY NC SA 4.0 (Annex A).

104. LA's Green New Deal. (2019). Sustainability plan. Available at https://plan. lamayor.org/sites/default/files/pLAn_2019_final.pdf.

105. Cochran, J., Denholm, P., Mooney, M., Steinberg, D., Hale, E., Heath, G., Palmintier, B., Sigrin, B., Keyser, D., McCamey, D., Cowiestoll, B., Horowitz, K., Horsey, H., Fontanini, A., Jain, H., Muratori, M., Jorgenson, J., Irish, M., Ban-Weiss, G., Cutler, H., Ravi, Vikram, N., Scott, C., Jaquelin, and Denholm, P. (2021). *The Los Angeles 100% Renewable Energy Study (LA100): Executive Summary*. Golden, CO, USA: National Renewable Energy Laboratory (NREL), NREL/TP-6A20-79444.

Available at https://doi.org/10.2172/1774871 and https://www.osti. gov/servlets/purl/1774871.

106. A full suite of EV charging infrastructure, operations, and cost analysis tools. Available at https://www.nrel.gov/transportation/evi-x.html.

107. U.S. Department of Energy. (2017). Confronting the duck curve: How to address over-generation of solar energy. Available at https://www.energy. gov/eere/articles/confronting-duck-curve-how-address-over-generation-solar-energy.

108. Arent, D. J., Barrows, C., Davis, S., *et al.* (2021). Integration of energy systems. *MRS Bulletin, 46,* 1139–1152. https://doi.org/10.1557/s43577-021-00244-8.

109. Bouma, A. T., Wei, Q. J., Parsons, J. E., Buongiorno, J., and Lienhard, J. H. (2022). Energy and water without carbon: Integrated desalination and nuclear power at Diablo Canyon. *Applied Energy*, 323, 119612. https:// doi.org/10.1016/j.apenergy.2022.119612.

110. https://flowcharts.llnl.gov/ for the U.S. and https://www.iea.org/sankey/.

111. IEA. (2023). *Energy Technology Perspectives 2023*. Paris: IEA. Available at https://www.iea.org/reports/energy-technology-perspectives-2023, License: CC BY 4.0.

112. Flores-Granobles, M. and Saeys, M. (2020). Minimizing CO_2 emissions with renewable energy: A comparative study of emerging technologies in the steel industry. *Energy Environmental Science*, 13, 1923–1932. https:// doi.org/10.1039/D0EE00787K.

113. Eicke, L. and De Blasio, N. (2022). The future of green hydrogen value chains: Geopolitical and market implications in the industrial sector. Belfer Center for Science and International Affairs, Harvard Kennedy School. Available at https://www.belfercenter.org/sites/default/files/files/ publication/Paper_MappingHydrogen_Final.pdf.

114. Sun, P., Young, B., Elgowainy, A., Lu, Z., Wang, M., Morelli, B., and Hawkins, T. (June 18 2019). Criteria air pollutants and greenhouse gas emissions from hydrogen production in U.S. steam methane reforming facilities. *Environmental Science & Technology,* 53(12), 7103–7113. https:// doi.org/10.1021/acs.est.8b06197.

115. Blain, L. Record-breaking hydrogen electrolyzer claims 95% efficiency. Available at https://newatlas.com/energy/hysata-efficient-hydrogen-electrolysis/ (Accessed August 2023).

116. Power, A. Neom green hydrogen project. Available at https://acwapower.com/en/projects/neom-green-hydrogen-project/.

117. IEA. (2022). *Global Hydrogen Review 2022*. Paris: IEA. Available at https://www.iea.org/reports/global-hydrogen-review-2022, License: CC BY 4.0.

118. Van de Graaf, T. (2022). *Hydrogen's Decade: The Global Race for Clean Hydrogen Means No Geopolitical Realities and Interdependence*. Finance and Development, International Monetary Fund (IMF).

119. Elishav, O., Lis, B. M., Miller, E. M., Arent, D. J., Valera-Medina, A., Dana, A. G., Shter, G. E., and Grader, G. S. (2020). Progress and prospective of nitrogen-based alternative fuels. *Chemical Reviews, 120*(12), 5352–5436. https://doi.org/10.1021/acs.chemrev.9b00538.

120. Ferdous, W., Bai, Y., Ngo, T. D., Manalo, A., and Mendis, P. (2019). New advancements, challenges and opportunities of multi-storey modular buildings — A state-of-the-art review. *Engineering Structures*, 183, 883–893, https://doi.org/10.1016/j.engstruct.2019.01.061.

121. Sustainable Forestry Initiative. Available at https://forests.org/.

122. Curtis, T. L., Buchanan, H., Smith, L., and Heath, G. (2021). *A Circular Economy for Solar Photovoltaic System Materials: Drivers, Barriers, Enablers, and U.S. Policy Considerations*. Golden, CO, USA: National Renewable Energy Laboratory (NREL), NREL/TP-6A20-74550. Available at https://www.nrel.gov.docs/fy21osti/74550. And PV Module Design for Recycling Guidelines available at https://iea-pvps.org/key-topics/pv-module-design-for-recycling-guidelines/.

123. Global Battery Alliance. Available at https://www.globalbattery.org/.

124. Burgess, M., Holmes, H., Sharmina, M., and Shaver, M. P. (2021). The future of UK plastics recycling: One bin to rule them all, resources. *Conservation and Recycling*, 164, and see The bio-optimized technologies to keep thermoplastics out of landfills and the environment consortium. Available at https://www.bottle.org/.

125. European Union. (n.d.). Circular economy action plan. Available at https://environment.ec.europa.eu/strategy/circular-economy-action-plan_en.

126. Platform for accelerating the circular economy. Available at https://pacecircular.org/.

127. Bordoff, J. Columbia University. See an extensive list of publications. Available at https://www.energypolicy.columbia.edu/jason-bordoff/.

128. Elkington, J. (2020). *Green Swans: The Coming Boom in Regenerative Capitalism.* New York: Fast Company Press.

129. IPCC AR6 WGII: IPCC, 2022: Summary for Policymakers [H.-O. Pörtner, D. C. Roberts, E. S. Poloczanska, K. Mintenbeck, M. Tignor, A. Alegría, M. Craig, S. Langsdorf, S. Löschke, V. Möller, and A. Okem (eds.)]. In: *Climate Change 2022: Impacts, Adaptation and Vulnerability. Contribution of Working Group II to the Sixth Assessment Report of the Intergovernmental Panel on Climate Change* [H.-O. Pörtner, D. C. Roberts, M. Tignor, E. S. Poloczanska, K. Mintenbeck, A. Alegría, M. Craig, S. Langsdorf, S. Löschke, V. Möller, A. Okem, B. Rama (eds.)]. Cambridge University Press, Cambridge, UK and New York, NY, USA, pp. 3–33, https://doi.org/10.1017/9781009325844.001.

130. United Nations Environment — Finance Initiative — Partnership between United Nations Environment and the Global Financial Sector to Promote Sustainable Finance. (n.d.). United Nations Environment — Finance Initiative — Partnership between United Nations Environment and the Global Financial Sector to promote sustainable finance. Available at https://www.unepfi.org/ (Accessed August 19 2023).

131. New York State. Department of Public Service. (n.d.). Reforming the energy vision. Available at https://www3.dps.ny.gov/w/pscweb.nsf/all/cc4f2efa3a23551585257dea007dcfe2.

132. Figure TS.2 in Pörtner, H.-O., Roberts, D. C., Adams, H., Adelekan, I., Adler, C., Adrian, R., Aldunce, P., Ali, E., Ara Begum, R., Bednar-Friedl, B., Bezner Kerr, R., Biesbroek, R., Birkmann, J., Bowen, K., Caretta, M. A., Carnicer, J., Castellanos, E., Cheong, T. S., Chow, W., Cissé, G., Clayton, S., Constable, A., Cooley, S., Costello, M. J., Craig, M., Cramer, W., Dawson, R., Dodman, D., Efitre, J., Garschagen, M., Gilmore, E. A., Glavovic, B., Gutzler, D., Haasnoot, M., Harper, S., Hasegawa, T., Hayward, B., Hicke, J. A., Hirabayashi, Y., Huang, C., Kalaba, K., Kiessling, W., Kitoh, A., Lasco, R., Lawrence, J., Lemos, M. F., Lempert, R., Lennard, C., Ley, D., Lissner, T., Liu, Q., Liwenga, E., Lluch-Cota, S., Löschke, S., Lucatello, S., Luo, Y., Mackey, B., Mintenbeck, K., Mirzabaev, A., Möller, V., Moncassim Vale, M., Morecroft, M. D., Mortsch, L., Mukherji, A., Mustonen, T., Mycoo, M., Nalau, J., New, M., Okem, A. (South Africa), Ometto, B.,

O'Neill, R., Pandey, C., Parmesan, M., Pelling, P. F., Pinho, J., Pinnegar, E. S., Poloczanska, A., Prakash, B., Preston, M.-F., Racault, D., Reckien, A., Revi, S. K., Rose, E. L. F., Schipper, D. N., Schmidt, D., Schoeman, R., Shaw, N. P., Simpson, C., Singh, W., Solecki, L., Stringer, E., Totin, C. H., Trisos, Y., Trisurat, M., van Aalst, D., Viner, M., Wairu, R., Warren, P., Wester, D., Wrathall, and Zaiton Ibrahim, Z. 2022: Technical Summary. [H.-O. Pörtner, D. C. Roberts, E. S. Poloczanska, K. Mintenbeck, M. Tignor, A. Alegría, M. Craig, S. Langsdorf, S. Löschke, V. Möller, and A. Okem (eds.)]. In: *Climate Change 2022: Impacts, Adaptation, and Vulnerability. Contribution of Working Group II to the Sixth Assessment Report of the Intergovernmental Panel on Climate Change* [H.-O. Pörtner, D.C. Roberts, M. Tignor, E. S. Poloczanska, K. Mintenbeck, A. Alegría, M. Craig, S. Langsdorf, S. Löschke, V. Möller, A. Okem, and B. Rama (eds.)]. Cambridge University Press, Cambridge, UK and New York, NY, USA, pp. 37–118, https://doi.org/10.1017/9781009325844.002.

133. European Commission. (2019). A European green deal. Available at https://commission.europa.eu/strategy-and-policy/priorities-2019-2024/european-green-deal_en.

134. U.S. Congress. (2022). H.R.4346 — Chips and science act. Available at https://www.congress.gov/bill/117th-congress/house-bill/4346.

135. Bordoff, J. (2022). America's landmark climate law. International Monetary Fund. Available at https://www.imf.org/en/Publications/fandd/issues/2022/12/america-landmark-climate-law-bordoff and https://time.com/6247230/inflation-reduction-act-global-response-climate-trade-protectionsim/.

136. World Economic Forum. (2022). Rethinking global supply chains for the energy transition. World Economic Forum. Available at https://www.weforum.org/agenda/2022/01/rethinking-supply-chains-for-the-energy-transition/.

137. *Economic Times*. (2023). Ladakh's renewable energy to get a boost with new transmission project. Available at https://energy.economictimes.indiatimes.com/news/renewable/ladakhs-renewable-energy-to-get-a-boost-with-new-transmission-project/97523404 and Government of India. (2022).

National green hydrogen mission. Available at https://www.india.gov.in/spotlight/national-green-hydrogen-mission.

138. Energy Foundation. (2022). China's 14th five-year plans on renewable energy development and modern energy system. Available at https://www.efchina.org/Blog-en/blog-20220905-en#:~:text=The%20plan%20targets%20a%2050,China's%20incremental%20electricity%20and%20energym.

139. Government of Canada. (2023). *A Made-in-Canada Plan: Affordable Energy, Good Jobs, and a Growing Clean Economy*. Available at https://www.budget.canada.ca/2023/report-rapport/chap3-en.html.

140. Government of Chile. (2021). Just energy transition strategy. Estrategia de Transición Justa en el sector Energía. Available at https://energia.gob.cl/sites/default/files/documentos/estrategia_transicion_justa_2021.pdf and National green hydrogen strategy. Available at https://energia.gob.cl/sites/default/files/national_green_hydrogen_strategy_-_chile.pdf.

141. Government of Australia. (2022). Transition to net zero. Available at https://www.globalaustralia.gov.au/industries/net-zero#:~:text=The%20Australian%20Government%20has%20made,and%20net%20zero%20by%202050.

142. Swim, J. K., Aviste, R., Lengieza, M. L., and Fasano, C. J. (2022). OK boomer: A decade of generational differences in feelings about climate change. *Global Environmental Change*, 73, 102479. https://doi.org/10.1016/j.gloenvcha.2022.102479.

143. The global energy alliance for people and planet. Available at https://www.energyalliance.org/.

144. Energy Transitions Commission. Available at https://www.energy-transitions.org/.

145. World Economic Forum: Climate change. Available at https://www.weforum.org/agenda/climate-change.

146. Clean Energy Ministerial. Available at https://www.cleanenergyministerial.org/.

147. IPCC, 2022: Summary for Policymakers [P. R. Shukla, J. Skea, A. Reisinger, R. Slade, R. Fradera, M. Pathak, A. Al Khourdajie, M. Belkacemi, R. van Diemen, A. Hasija, G. Lisboa, S. Luz, J. Malley, D. McCollum, S. Some, and P. Vyas, (eds.)]. In: *Climate Change 2022: Mitigation of*

Climate Change. Contribution of Working Group III to the Sixth Assessment Report of the Intergovernmental Panel on Climate Change [P. R. Shukla, J. Skea, R. Slade, A. Al Khourdajie, R. van Diemen, D. McCollum, M. Pathak, S. Some, P. Vyas, R. Fradera, M. Belkacemi, A. Hasija, G. Lisboa, S. Luz, and J. Malley, (eds.)]. Cambridge University Press, Cambridge, UK and New York, NY, USA, https://doi.org/10.1017/9781009157926.001.

148. Arent, D. J., Pless, J., and Statwick, P. (2016). *Five Forces of 21st Century Innovation Strategy: Insights for Leaders*. Golden, CO, USA: National Renewable Energy Laboratory (NREL). https://www.osti.gov/servlets/purl/1244311.

149. https://www.cop28.com/en/global-renewables-and-energy-efficiency-pledge (Accessed February 5 2024).

Index